# Soil Testing & Fertiliser Recommendations

# Soil Testing & Fertiliser Recommendations

Editor

**Surendra Mohite**

scitus
academics

**Soil Testing & Fertiliser Recommendations**

Edited by **Surendra Mohite**

ISBN: 978-1-68117-097-8

Library of Congress Control Number: 2015935497


SCITUS Academics LLC,
www.scitusacademics.com
Box No. 4766, 616 Corporate Way,
Suite 2, Valley Cottage,
NY 10989



Printed in United States of America on Acid Free Paper ∞

# Contents

# Preface

In agriculture, a soil test is the analysis of a soil sample to determine nutrient and contaminated content, composition, and other characteristics such as the acidity or pH level. A soil test can determine fertility, or the expected growth potential of the soil which indicates nutrient deficiencies, potential toxicities from excessive fertility and inhibitions from the presence of non-essential trace minerals. The test is used to mimic the function of roots to assimilate minerals. The expected rate of growth is modelled by the Law of the Maximum. Soil testing is used to facilitate fertilizer composition and dosage selection for land employed in both agricultural and horticultural industries. Lab tests are more accurate, though both types are useful. In addition, lab tests frequently include professional interpretation of results and recommendations. Always refer to all proviso statements included in a lab report as they may outline any anomalies, exceptions, and shortcomings in the sampling and/or analytical process/results. A fertilizer is any material of natural or synthetic origin (other than liming materials) that is applied to soils or to plant tissues (usually leaves) to supply one or more plant nutrients essential to the growth of plants. Conservative estimates report 30 to 50% of crop yields are attributed to natural or synthetic commercial fertilizer.

**Editor**

Chapter 1

# Reducing Runoff Loss of Applied Nutrients in Oil Palm Cultivation Using Controlled-Release Fertilizers

A. Bah[1], M. H. A. Husni[1], C. B. S. Teh[1], M. Y. Rafii[2], S. R. Syed Omar[3], and O. H. Ahmed[4]

[1]Department of Land Management, Faculty of Agriculture, Universiti Putra Malaysia, 43400 Serdang, Selangor, Malaysia

[2]Institute of Tropical Agriculture, Universiti Putra Malaysia, 43400 Serdang, Selangor, Malaysia

[3]Diversatech (M) Fertilizer Sdn. Bhd., Bandar Baru Bangi, 43650 Selangor, Malaysia

[4]Department of Crop Science, Faculty of Agriculture and Food Sciences, Universiti Putra Malaysia, Bintulu Campus Sarawak, 97008 Bintulu, Sarawak, Malaysia

# ABSTRACT

Controlled-release fertilizers are expected to minimize nutrient loss from crop fields due to their potential to supply plant-available nutrients in synchrony with crop requirements. The evaluation of the efficiency of these fertilizers in tropical oil palm agro ecological conditions is not yet fully explored. In this study, a one-year field trial was conducted to determine the impact of fertilization with water soluble conventional mixture and controlled-release fertilizers on runoff loss of nutrients from an immature oil palm field. Soil and nutrient loss were monitored for one year in 2012/2013 under erosion plots of 16 $m^2$ on 10% slope gradient. Mean sediments concentration in runoff amounted to about 6.41 $t\,ha^{-1}$. Conventional mixture fertilizer posed the greatest risk of nutrient loss in runoff following fertilization due to elevated nitrogen (6.97%), potassium (13.37%), and magnesium (14.76%) as percentage of applied nutrients. In contrast, this risk decreased with the application of controlled-release fertilizers, representing 0.75–2.44% N, 3.55–5.09% K, and 4.35–5.43% Mg loss. Meanwhile, nutrient loss via eroded sediments was minimal compared with loss through runoff. This research demonstrates that the addition of controlled-release fertilizers reduced the runoff risks of nutrient loss possibly due to their slow-release properties.

# INTRODUCTION

Oil palm is mainly cultivated on highly weathered soils which belong to the orders Ultisols and Oxisols. These soils are predominantly acidic and low in fertility [1]. Fertilizers are crucial in oil palm production, accounting for 50–70% of field operational costs and about 25% of the total cost of production [2, 3]. Mineral fertilizers, mainly conventional forms, account for more than 90% of fertilizers used by all types of farming systems in Malaysia [4]. The oil palm is a heavy feeder and requires quite large quantities of fertilizers to produce good yield [5]. Fertilizer management on undulating, hilly soils used for oil palm cultivation is very important because of the need to maintain fertility of the soil and to as well minimize soil erosion and nutrient loss. Frequent application of large amounts of chemical fertilizers coupled with high rainfall intensity tends to increase the risk of nutrient loss. The loss of nutrients through leaching and runoff reduces both crop productivity and economic gains. Furthermore, excess nutrient loading to ground or underground water bodies can impair designated uses of water [6, 7]. There is a need to develop alternatives from the fertilizer industry to make mineral fertilization for high value crops such as oil palm more economically viable and ecologically compatible.

Approaches to improve crop nutrient use efficiency have been proposed [8, 9]. Among the management practices to both improve fertilizer efficiency and reduce environmental pollution, the use of slow-release fertilizers (SRFs) and controlled-release fertilizers (CRFs) seems to be promising for widespread use in agriculture [10–13]. As compared to conventional fertilizers, the gradual release of nutrients from CRFs could be synchronized with plant needs and minimize nutrient loss through runoff and leaching to ultimately improve fertilizer use efficiency. According to the International Fertilizer Industry Association [14], controlled-release nitrogen fertilizers have agronomic advantages, especially in the tropics and in regions with light-textured soils and under heavy rainfall or irrigation, where N losses are particularly high.

Presently, Malaysia's oil palm industry is faced with a growing challenge of labor shortage partly due to frequent application of straight fertilizers in 4–6 splits per annum. Therefore, one of the possible approaches to address this challenge is through the adoption of improved fertilizer technologies such as the use of CRFs where fertilizer application rounds can be reduced to two splits per annum. The evaluation of efficiency of CRF is necessary to help oil palm plantations make informed decisions. The objective of this study was to evaluate whether application of controlled-release fertilizers (AJIB CRF) instead of water soluble conventional mixture fertilizer can reduce the risk of nutrient loss via runoff and erosion.

# MATERIALS AND METHODS

## Experimental Site Details

The study was carried out in 2012/2013 on a newly established oil palm farm at the experimental station of University Putra Malaysia

Agriculture Park in Puchong, Selangor (02°N 59.035′, 101°E 38.913′). Initially, the field was a fallow grassland. The area has a humid tropical climate with a mean annual rainfall of 2700 mm and temperature of 25.3°C [15]. Experimental plots measuring 4 m × 4 m and delineated along uniform land slope of 10% were demarcated with transparent plastic sheets inserted 5 cm deep into the soil and 15 cm above soil surface to prevent lateral flow and control the risk of plot contamination (Figure 1). Twelve-month-old oil palm clones (AA Hybrida IS) obtained from Applied Agricultural Resources Company were planted using the 9 m × 9 m × 9 m triangular system. Daily precipitation records were taken throughout the experimental period using rain gauge. Runoff was routed via a V-shaped stainless steel aluminum spout attached to a funnel-fitted tank. The experiment was arranged in three blocks (replicates) with six treatments established perpendicular to the slope.

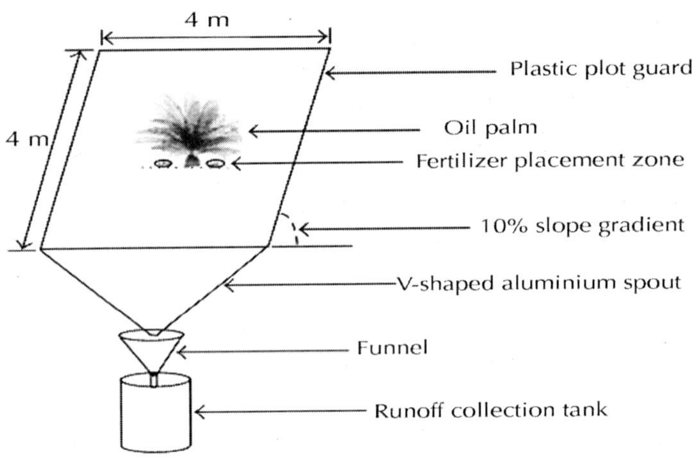


**Figure 1:** Experimental plot setup.

## Experimental Soil

Soil particle size distribution measurement was based on the pipette method [16], while the bulk density of the soil was determined using core rings as outlined by Teh and Talib [17]. Soil pH, EC, and CEC were determined following standard protocols [18–20]. Total soil carbon and nitrogen were analyzed using TruMac CNS Analyzer (LECO, St.

Joseph), which utilizes the dry combustion principle. Soil exchangeable K, Ca, Mg, and CEC using 1 M $NH_4OAc$ (pH 7.0) [21, 22] and P using Bray II method [23] were determined.

## Fertilizer Treatments

The treatments evaluated in this study were as follows:

- control (no fertilizer),
- mixture (water soluble mixed fertilizer),
- CRFB-60% (60% dosage of briquette CRF),
- CRFG-60% (60% dosage of granular CRF),
- CRFB-100% (full dosage of briquette CRF),
- CRFG-100% (full dosage of granular CRF).

T2 consists of a mixture of water soluble straight fertilizers, namely, ammonium sulfate (21% N), Christmas Island Rock Phosphate (33% $P_2O_5$), Muriate of Potash (60% $K_2O$), and Kieserite (27% MgO). The grade of the coated CRFs (AJIB CRF) is 10-6-20-2. Briquette CRF is spherical in shape with a size of about 5 cm in length, while granular CRF is pelletized and is approximately 1 cm in length. The full dosage of fertilizer application was based on standard recommended rates for immature oil palm. The treatments were arranged in a randomized complete block design (RCBD) with three replications (Table 1). The fertilizers were surface-placed under the plant canopy in two splits per year at six months' intervals.

**Table 1:** Fertilizer application treatments

| **Fertilizer treatments** | **Nutrient rate (kg ha$^{-1}$ yr$^{-1}$)** | | | |
|---|---|---|---|---|
| | **N** | **P** | **K** | **Mg** |
| Control | 0.00 | 0.00 | 0.00 | 0.00 |
| Mixture | 48.96 | 13.07 | 81.60 | 5.86 |
| CRFB-60% | 32.64 | 8.71 | 54.40 | 3.91 |
| CRFG-60% | 32.64 | 8.71 | 54.40 | 3.91 |
| CRFB-100% | 48.96 | 13.07 | 81.60 | 5.86 |
| CRFG-100% | 48.96 | 13.07 | 81.60 | 5.86 |

## Runoff Collection and Analysis

Following each rainfall event, surface runoff water plus eroded sediments per plot was collected for one year (October 2012 to September 2013) using tanks. For each rainfall event, the total water volume per each tank was recorded and 1 L subsamples were collected in detergent-washed polypropylene bottles, after which they were transported to the laboratory, filtered, and analyzed for N, P, K, and Mg content. All tanks were emptied and thoroughly cleaned with distilled water after each sample collection event. The sediments retained after filtration of runoff samples with Whatman number 5 were oven-dried at 60°C for 24 h and weighed with an electronic balance. Sediment loss from each plot was calculated by multiplying the volume of runoff with the nutrients concentration in the analyte. Soil loss was converted into sediment yield in $t\,ha^{-1}$. Dried sediments and runoff filtrates were separately analyzed for their N, P, K, and Mg contents following standard procedures as stated above.

Analysis of variance was used to detect treatment effect, whereas treatment means were compared using Duncan's new multiple range test (DNMRT) at ($P \leq 0.05$) with statistical analysis system (SAS) software (Version 9.3).

# RESULTS AND DISCUSSION

## Rainfall, Runoff, and Soil Loss

The texture of the soil in the study area is classified as sandy clay loam (Typic Kandiudult) having a deep kandic horizon with clay content between 18 and 35% [24]. Based on texture and its friable nature, this soil is prone to erosion but may also be highly permeable to enhance water infiltration. Although soil consisted of 63% sand, clay 31%, and 6% silt, there was substantial loss of silt (40%) in the eroded sediments compared to sand (32%) and clay (28%) as shown in Table 2. The higher loss of silt particles in the eroded sediments relates to the lightness of this particle.

**Table 2:** Soil textural classification of study site (0–20 cm) and sediments distribution in runoff

| Sample | Textural distribution (%) | | | Soil texture class (USDA) |
|---|---|---|---|---|
| | Sand | Silt | Clay | |
| Soil | 63 | 6 | 31 | Sandy clay loam |
| Sediments | 32 | 40 | 28 | Clay loam |

The annual recorded precipitation was 2635 mm. Highest observed rainfall occurred in the months of October (316 mm) and November (302 mm) in 2012. A short dry spell was observed in February, June, and August, 2013, as a result of seasonal changes in rainfall distribution patterns in the area. Figure 2 shows monthly generated runoff and soil loss due at the experimental site. A total of 371 mm of runoff volume was recorded during the study period. Also, runoff volume as percentage of rainfall ranged from 5 to 20% with a mean value of 14%. This result relates to the relatively high infiltration (5.40 mm $hr^{-1}$) of the soil. The annual sediment loss was 6.41 t $ha^{-1}$ and maximum loss (1.05 t $ha^{-1}$) occurred in the month of November 2012 (Figure 2).

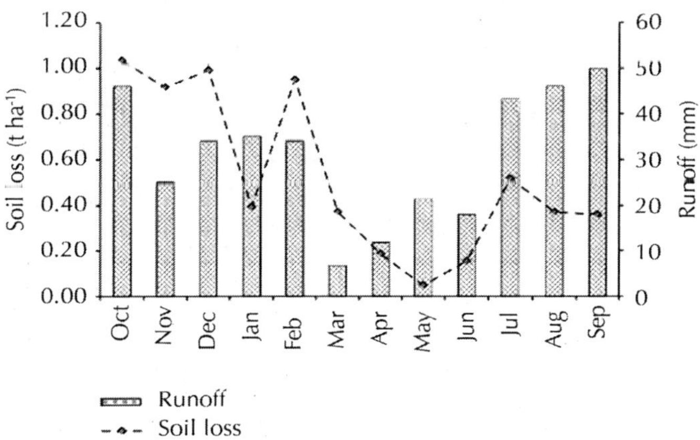


**Figure 2:** Monthly soil loss and runoff during study period (2012/2013).

Soil erosion studies in oil palm fields in Malaysia have shown a wide range of soil loss. According to reports by PORIM [25] and Harte mink [26] soil loss ranges from 13 to 78 Mg $ha^{-1}$ $yr^{-1}$ for Oxisols and 1 to 28 Mg $ha^{-1}$ $yr^{-1}$ for Ultisols. The results of the present study are comparable to that of Corley and Tinker [27] who reported soil loss of less than 9 t $ha^{-1}$. Similar findings have also been previously reported [28, 29], where soil erosion losses from oil palm fields were in the range of 7.47–14.92 t $ha^{-1}$. Loss of nutrients through runoff and sediments relates to soil texture, age of palms, soil topography and infiltrability, land clearing, inadequate soil conservation measures, and the lag time between fertilizer application and rainfall [30, 31].

## Nutrient Loss in Runoff and Eroded Sediments

As shown in Table 3, net annual losses of nutrients in runoff varied significantly between soluble mixture and CRF treatments, but the differences were not significant between briquette and granular CRFs. Comparison of the overall nutrients concentration of the different treatments indicated that runoff losses of nutrients were highest in mixture fertilizer compared with the controlled-release fertilizers (AJIB CRF) and the control. The loss of nutrients from applied mixture treatments was rapid at the initial stage following first fertilization in October, after which it declined rapidly and picked up again in June following second fertilization (Figure 3). Cumulatively, the annual loss of N through runoff in the plots treated with mixture fertilizers accounted was 4.78 kg $ha^{-1}$ whereas the N loss from applied CRF fertilizers ranged between 1.73 and 2.56 kg $ha^{-1}$. The net losses of K and Mg were greatest for mixture fertilizer treatment, accounting for 13% and 15%, respectively, of the added fertilizer. The average annual loss of P in the runoff due to fertilization was relatively low with full dosage of granular CRF accounting for the highest loss (4%). The losses of nutrients from the reduced dosages of CRFs (CRFB-60% and CRFG-60%) were consistently lower than their full dosage counterparts, although the latter showed better plant response in terms of N and Mg uptake. By comparing the nutrients lost through surface runoff with those lost in the eroded sediments (Table 3), it is clear that runoff losses of nutrients in all treatments were substantially greater than those lost in displaced sediments.

**Table 3:** Summary of net nutrient loss in oil palm ecosystem through surface runoff and eroded sediments

| Fertilizer treatments | Net loss in runoff (kg ha$^{-1}$ yr$^{-1}$) | Net loss in eroded sediments (kg ha$^{-1}$ yr$^{-1}$) | Net total loss (kg ha$^{-1}$ yr$^{-1}$) | **Net loss as % of nutrient applied** |
|---|---|---|---|---|
| | Nitrogen (N) | | | |
| Control | $1.20^{c} \pm 0.04$ | $0.17^{c} \pm 0.01$ | $1.37^{c}$ | — |
| Mixture | $3.85^{a} \pm 0.22$ | $0.93^{a} \pm 0.27$ | $4.78^{a}$ | 6.97 |
| CRFB-60% | $1.33^{bc} \pm 0.02$ | $0.40^{ab} \pm 0.24$ | $1.73^{bc}$ | 0.75 |
| CRFG-60% | $1.39^{bc} \pm 0.24$ | $0.39^{ab} \pm 0.20$ | $1.78^{bc}$ | 0.84 |
| CRFB-100% | $1.81^{b} \pm 0.26$ | $0.41^{ab} \pm 0.12$ | $2.22^{b}$ | 1.74 |
| CRFG-100% | $1.98^{b} \pm 0.30$ | $0.58^{ab} \pm 0.16$ | $2.56^{b}$ | 2.44 |
| | Phosphorus (P) | | | |
| Control | $0.40^{cb} \pm 0.16$ | $0.12^{b} \pm 0.01$ | $0.52^{b}$ | — |
| Mixture | $0.70^{ab} \pm 0.09$ | $0.30^{a} \pm 0.07$ | $1.00^{a}$ | 3.74 |
| CRFB-60% | $0.45^{b} \pm 0.14$ | $0.24^{ab} \pm 0.05$ | $0.69^{ab}$ | 1.31 |
| CRFG-60% | $0.49^{b} \pm 0.01$ | $0.29^{a} \pm 0.03$ | $0.78^{ab}$ | 2.02 |
| CRFB-100% | $0.64^{ab} \pm 0.03$ | $0.31^{a} \pm 0.02$ | $0.95^{a}$ | 3.28 |
| CRFG-100% | $0.92^{a} \pm 0.19$ | $0.15^{b} \pm 0.06$ | $1.07^{a}$ | 4.22 |
| | Potassium (K) | | | |
| Control | $3.70^{c} \pm 0.17$ | $0.45^{c} \pm 0.11$ | $3.70^{c}$ | — |
| Mixture | $12.48^{a} \pm 0.87$ | $2.14^{a} \pm 0.37$ | $14.62^{a}$ | 13.37 |
| CRFB-60% | $5.56^{b} \pm 0.13$ | $1.08^{b} \pm 0.21$ | $6.64^{b}$ | 3.60 |
| CRFG-60% | $5.35^{b} \pm 0.17$ | $1.25^{b} \pm 0.37$ | $6.60^{b}$ | 3.55 |
| CRFB-100% | $5.96^{b} \pm 0.16$ | $1.32^{b} \pm 0.17$ | $7.28^{b}$ | 4.38 |
| CRFG 100% | $6.83^{b} \pm 0.35$ | $1.03^{b} \pm 0.23$ | $7.86^{b}$ | 5.09 |
| | Magnesium (Mg) | | | |
| Control | $0.51^{b} \pm 0.04$ | $0.11^{c} \pm 0.02$ | $0.62^{c}$ | — |
| Mixture | $1.20^{a} \pm 0.14$ | $0.28^{a} \pm 0.05$ | $1.48^{a}$ | 14.76 |
| CRFB-60% | $0.70^{ab} \pm 0.08$ | $0.19^{ab} \pm 0.04$ | $0.89^{b}$ | 4.72 |
| CRFG-60% | $0.58^{b} \pm 0.03$ | $0.29^{a} \pm 0.10$ | $0.87^{b}$ | 4.35 |
| CRFB-100% | $0.76^{ab} \pm 0.07$ | $0.17^{ab} \pm 0.02$ | $0.93^{b}$ | 5.43 |
| CRFG-100% | $0.68^{ab} \pm 0.02$ | $0.21^{ab} \pm 0.09$ | $0.89^{b}$ | 4.65 |

All analyses are mean ± standard error of mean (SEM). Means not sharing a common letter are significantly different by DNMRT ($P \leq 0.05$).

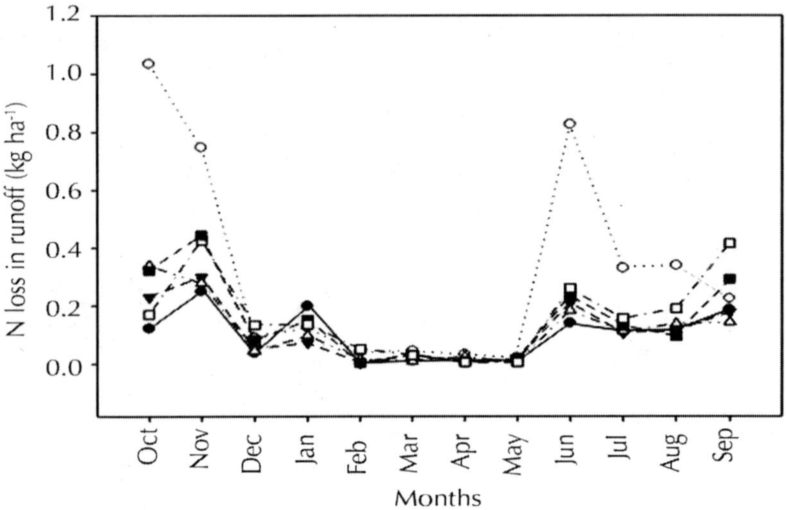
1.2
1.0
0.8
0.6
0.4
0.2
0.0
N loss in runoff (kg ha-1)
Oct
Nov
Dec
Jan
Feb
Mar
Apr
May
Jun
Jul
Aug
Sep
Months

(a)

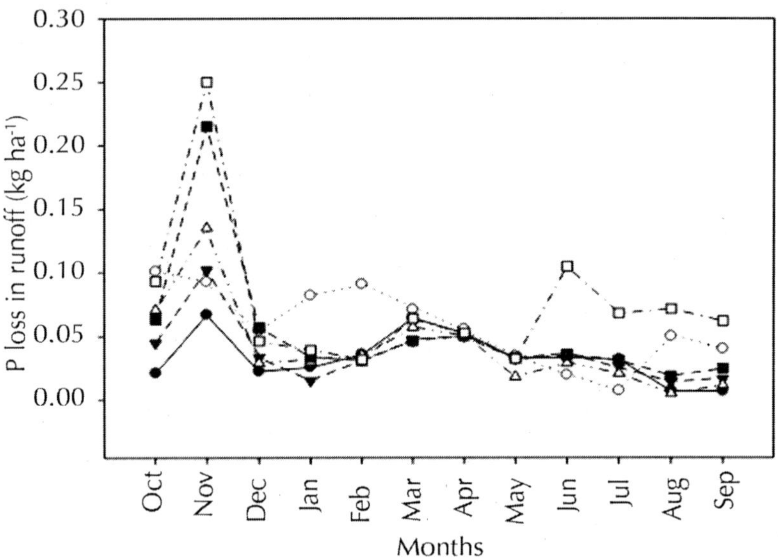
0.30
0.25
0.20
0.15
0.10
0.05
0.00
P loss in runoff (kg ha-1)
Oct
Nov
Dec
Jan
Feb
Mar
Apr
May
Jun
Jul
Aug
Sep
Months

(b)

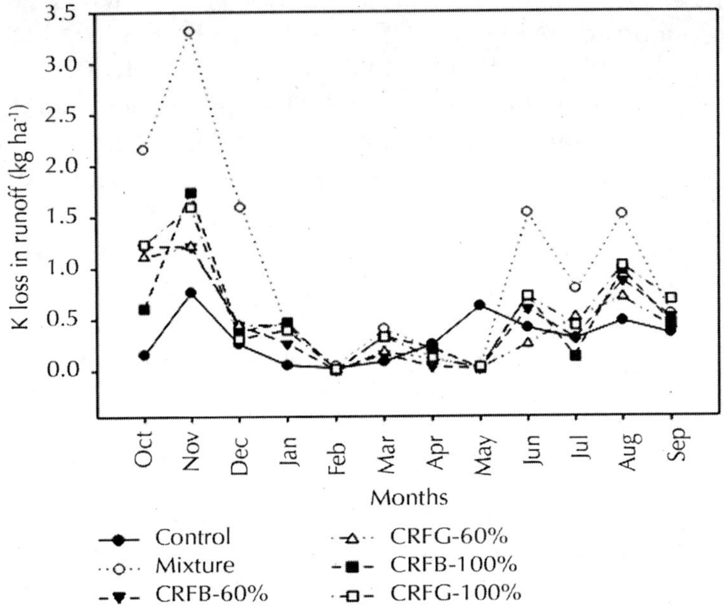


(c)

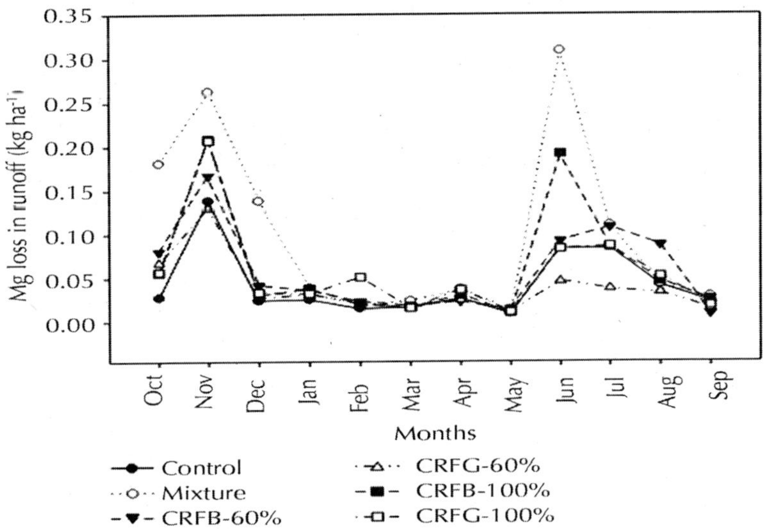


**(d)**

**Figure 3:** Monthly runoff loss patterns of nutrients as impacted by fertilizer application.

Nutrient loss data in this current study are consistent with other studies conducted in Malaysia. For example, studies conducted by Maene et al. [28] showed that 11% N, 3% P, 5% K, 6% Mg, and 5% Ca of applied fertilizers were lost through surface runoff alone during a low rainfall (1426 mm) on a 9% slope. The findings further indicated that the harvesting paths are the most susceptible areas to surface runoff due to compaction of the soil by machinery. Another study [32] revealed that 5–8% N, 10–15% K, 4–6% Mg, and <2% P were lost through runoff. This suggests that soluble nutrients such as N, K, and Mg are more susceptible to runoff losses. According to Wallace et al. [33], large losses of nutrients via surface runoff are still possible when a large rainfall event occurs soon after application of a fertilizer amendment.

The macronutrient release pattern varied substantially between mixture and CRF fertilizers (Figure 3). The nutrients released from mixture fertilizer (except for P) were higher and varied more after fertilization. Coated fertilizers (CRFs) gradually released nutrients compared to that of the mixture fertilizer. Mixture fertilizer showed lower P release in runoff due to low solubility of its P content from the phosphate rock source in contrast with CRF treatments, which are formulated with diammonium phosphate (DAP) as P source.

An examination of the nutrient release pattern shows that, immediately after fertilizer application in October 2012 and April 2013, the mean soluble N, K, and Mg concentrations in runoff were greatest for mixture fertilizer treatment and then followed by CRF treatments (Figure 3). Mean loss of nutrients in runoff from fertilized plots decreased significantly from the months of January to May, such that losses were similar to those observed for the unfertilized treatment. This trend is more obvious for nutrients such as N, K, and Mg than for P.

## Effect of Rainfall Intensity

The effects of rainfall intensity on nutrient loss were explored for the months of November (2012) and April (2013) and the findings suggest that 76% of the rainfall occurred at intensities below 10 mm $hr^{-1}$ and 13% in excess of 20 mm $hr^{-1}$ (Figure 4). Research findings suggest that rainfall intensity is one of the most important factors that contribute to soil loss and runoff [34, 35]. Soil erosion due to rainfall involves series of complex processes commencing from detachment by raindrop impact, transport of entrained soil particles by rain splash or surface flow, and then deposition. There is strong relationship between rainfall, runoff, and soil erosion by water. The occurrence of runoff, whether small or huge, always results in decline of soil fertility due to loss of topsoil and nutrients, loss of organic matter, and the consequent loss of the soil's capacity to retain nutrients and water.

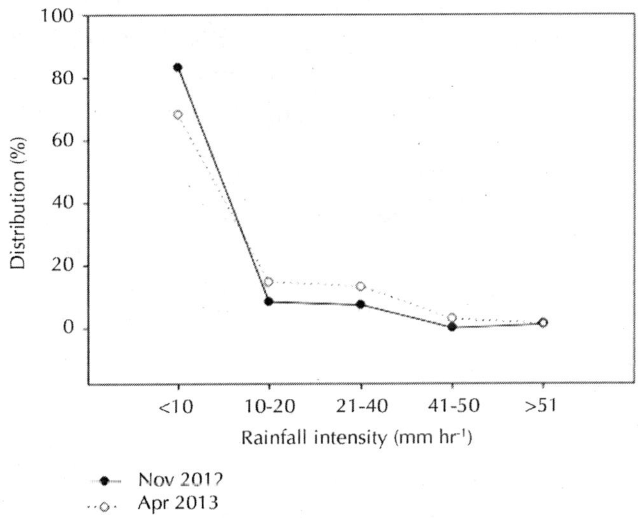


**Figure 4:** Rainfall intensity distribution for the months of November (2012) and April (2013).

Nutrient runoff potential during storm events is influenced by numerous factors, including amount and intensity of rainfall, antecedent rainfall conditions, timing and rate of fertilization, ponding, and irrigation management practices [36]. Rainfall intensity is considered a key determinant factor that influences the loss of soil and nutrients. As shown in Figure 4, the loss of soil through runoff significantly increased with increase in rainfall intensity. A recent study on the interactions between rainfall intensities and soil surface interrill erosion [37] showed that the soil erosion rates spiked following sharp increase in rainfall intensity. Attempts were conducted to evaluate P and N in surface runoff in relation to rainfall intensity and hydrology for two soils along a single hill slope and the results revealed that nutrient loss was significantly greater under the high intensity rainfall due to larger runoff volumes [38].

Linear relationships were observed between the rainfall intensity and nutrient (N) loss (Figure 5). The coefficients of determination $(R)^2$ of the linear regressions were high () for all treatments, ranging from 0.778 to 0.939, indicating the influence of rainfall intensity on nutrient loss. A study on nutrient loss of a limed soil in a reservoir area found a positive regressive relationship between nutrient loss and rainfall intensity [39]. Previously, it has also been found that the loss of P has a strongly positive relationship with rainfall intensity [40]. A study on the effect of rainfall intensity and soil surface cover on losses of sediment and soil organic carbon (SOC) via surface runoff found that higher rainfall intensity and lower vegetative cover produced higher sediment and consequently higher nutrient loss [41].

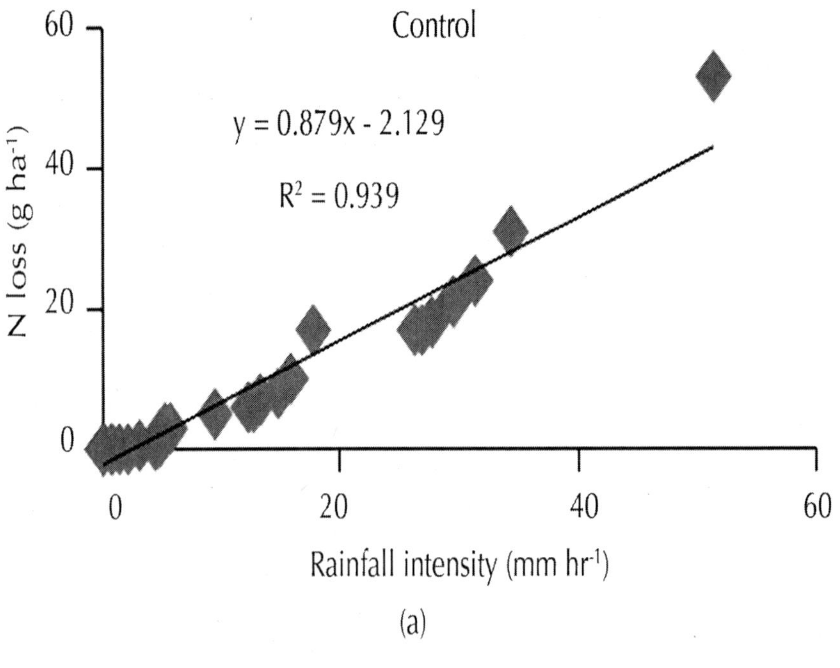


(a)

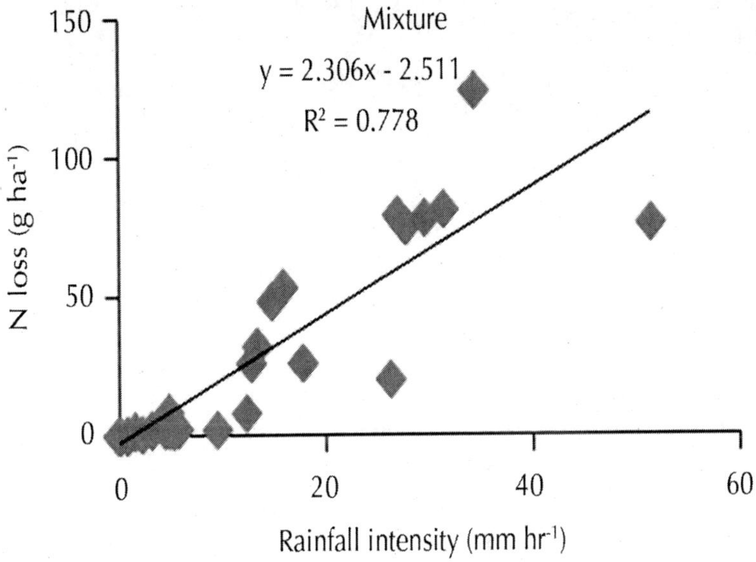
Mixture
y = 2.306x - 2.511
R² = 0.778
150
100
50
0
N loss (g ha⁻¹)
0
20
40
60
Rainfall intensity (mm hr⁻¹)

(b)

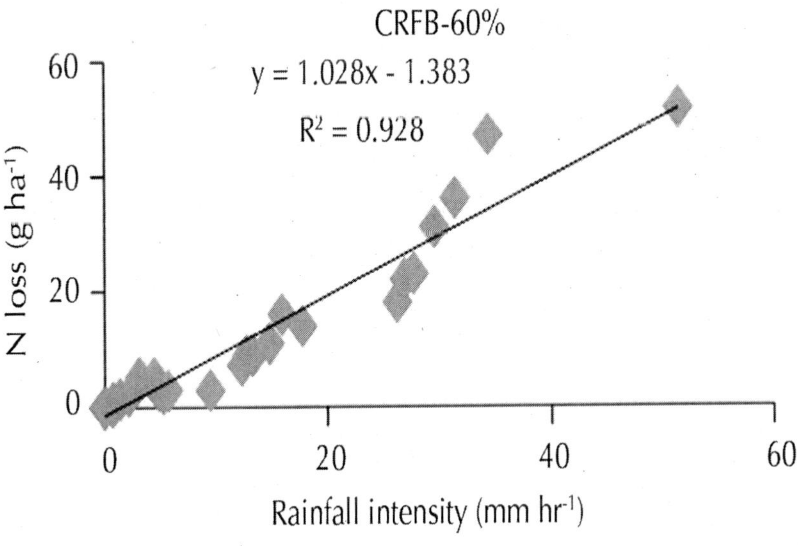
CRFB-60%
y = 1.028x - 1.383
R² = 0.928
60
40
20
0
N loss (g ha⁻¹)
0
20
40
60
Rainfall intensity (mm hr⁻¹)

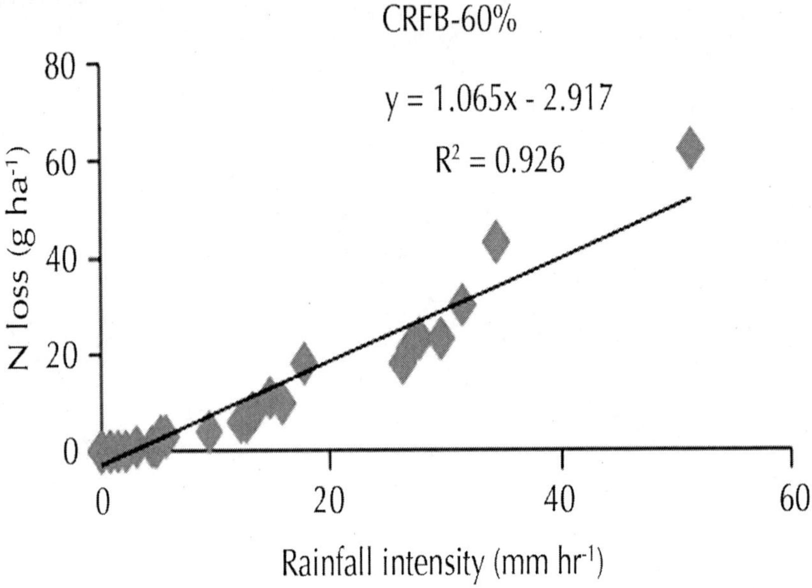
CRFB-60%
y = 1.065x - 2.917
R² = 0.926
N loss (g ha⁻¹)
0
20
40
60
80
0
20
40
60
Rainfall intensity (mm hr⁻¹)

(d)

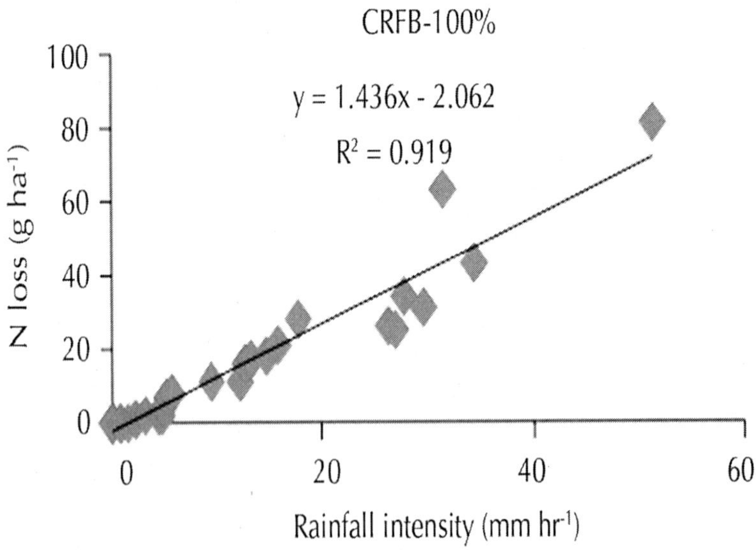
CRFB-100%
y = 1.436x - 2.062
R² = 0.919
N loss (g ha⁻¹)
0
20
40
60
80
100
0
20
40
60
Rainfall intensity (mm hr⁻¹)

(e)

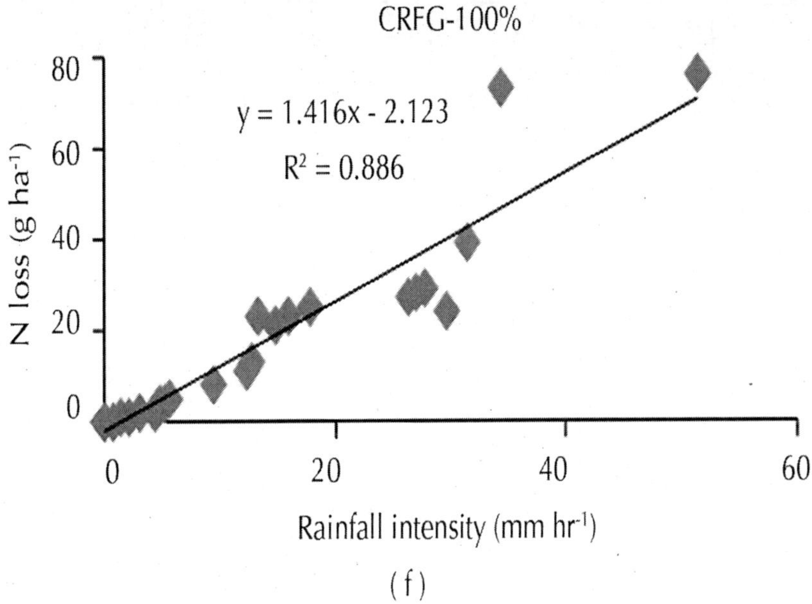


**Figure 5:** Relationship between rainfall intensity and nutrient (N) loss for different fertilizer treatments following each rainfall event.

## Oil Palm Response to Fertilizer Treatments

Oil palm response to fertilizer treatments during the first six months after fertilization (MAF) appears to show no significant variation ($p \leq 0.05$) in terms of frond production. However, at 12 MAF, plants that received treatments of granular CRFs showed significant improvement in frond number by 64–66% compared to mixture (59%) and control (52%) treatments (Table 4).

**Table 4:** Frond number of immature oil palms at different months after fertilization (MAF)

| Fertilizer treatments | Frond number plant$^{-1}$ | | | | |
|---|---|---|---|---|---|
| | 0 MAF | 3 MAF | 6 MAF | 9 MAF | 12 MAF |
| Control | $13^{a} \pm 0.12$ | $17^{c} \pm 0.50$ | $20^{b} \pm 0.49$ | $23^{c} \pm 0.99$ | $26^{c} \pm 1.53$ |
| Mixture | $13^{a} \pm 0.29$ | $19^{ab} \pm 1.15$ | $24^{ab} \pm 1.00$ | $28^{b} \pm 0.55$ | $32^{b} \pm 0.75$ |
| CRFB-60% | $12^{a} \pm 0.27$ | $19^{ab} \pm 0.27$ | $26^{a} \pm 0.82$ | $29^{ab} \pm 0.56$ | $33^{ab} \pm 0.67$ |
| CRFG-60% | $12^{a} \pm 0.27$ | $18^{ab} \pm 0.81$ | $27^{a} \pm 0.46$ | $31^{a} \pm 0.22$ | $35^{a} \pm 1.20$ |

| | | | | | |
|---|---|---|---|---|---|
| CRFB-100% | $12^{a} \pm 0.12$ | $19^{ab} \pm 0.87$ | $25^{ab} \pm 1.35$ | $31^{a} \pm 1.05$ | $34^{ab} \pm 1.33$ |
| CRFG-100% | $13^{a} \pm 0.24$ | $20^{a} \pm 0.43$ | $26^{a} \pm 1.98$ | $32^{a} \pm 0.66$ | $36^{a} \pm 0.48$ |

All analyses are mean ± standard error of mean (SEM). Means not sharing a common letter are significantly different by DNMRT (P≤0.05).

The effects of fertilizer treatments on foliar nutrient content of frond #3 are given. In general, leaf nutrient concentrations for N, P, K, and Mg for fertilized plants were higher at 12 MAF compared to 0, 3, 6, and 9 MAF. This is probably due to the cumulative effect of the fertilizers which were applied in two rounds per annum at six months' interval. Plant response in terms of nutrients uptake appears to be higher with application of full dose CRF treatments (Table 5). For full dose granular CRF treated plants, foliar N increased from 2.10% to 2.46%, while foliar K and Mg increased from 1.35% to 1.75% and 0.25% to 0.55%, respectively. Leaf analyses for the purpose of diagnosing nutritional status of oil palm have proven successful in most industrial plantations worldwide for several decades, due to the fact that the obtained information serves as a reliable indicator for annual guidance of mineral fertilization [42]. It has been reported that foliar concentrations of N, Mg, B, Cu, Fe, and Mo were significantly affected by controlled-release fertilizer treatments on container-grown ponderosa pine seedlings [43].

**Table 5:** Effects of fertilizer treatments on leaf and rachis nutrients concentration of oil palm at 12 MAF

| Fertilizer treatments | Leaf nutrients concentration (%) | | | |
|---|---|---|---|---|
| | **N** | **P** | **K** | **Mg** |
| Control | $2.03^{d} \pm 0.03$ | $0.12^{c} \pm 0.01$ | $1.10^{d} \pm 0.01$ | $0.26^{c} \pm 0.02$ |
| Mixture | $2.28^{bc} \pm 0.03$ | $0.20^{ab} \pm 0.02$ | $1.55^{bc} \pm 0.01$ | $0.42^{b} \pm 0.01$ |
| CRFB-60% | $2.27^{ab} \pm 0.13$ | $0.21^{ab} \pm 0.01$ | $1.57^{bc} \pm 0.04$ | $0.43^{b} \pm 0.02$ |
| CRFG-60% | $2.30^{bc} \pm 0.09$ | $0.20^{ab} \pm 0.01$ | $1.53^{c} \pm 0.03$ | $0.44^{b} \pm 0.01$ |
| CRFB-100% | $2.38^{a} \pm 0.10$ | $0.24^{a} \pm 0.01$ | $1.66^{ab} \pm 0.01$ | $0.56^{a} \pm 0.02$ |
| CRFG-100% | $2.46^{a} \pm 0.03$ | $0.25^{a} \pm 0.01$ | $1.75^{a} \pm 0.04$ | $0.55^{a} \pm 0.01$ |

All analyses are mean ± standard error of mean (SEM). Means not sharing a common letter are significantly different by DNMRT (P≤0.05).

# CONCLUSIONS

Controlled-release fertilizers (AJIB CRF) application in an immature oil palm field potentially decreased surface runoff loss of nutrients. This could be attributed to the fact that nutrient elements in CRFs are readily and slowly available for plant uptake over a given period. It is imperative to note that fertilizers should be applied during periods of less rainfall events in order to limit nutrient and soil loss due to the effect of high rainfall intensity. Considering the long term economic viability and environmental factors associated with nutrient loss, oil palm plantations may consider application of CRFs as an improved fertilizer management option. More research is needed to better elucidate mechanism of nutrient loss from oil palm ecosystems during specific storm events.

# ACKNOWLEDGMENTS

The authors wish to acknowledge Diversatech (M) Fertilizer Sdn. Bhd., Universiti Putra Malaysia, and Commonwealth Scholarship and Fellowship Plan (CSFP) through Ministry of Education Malaysia (MOE) for financial and technical support. Thanks are due to Madam or Asma Mohd Zaki for her help in the laboratory work.

# REFERENCES

1. J. Shamshuddin and M. Anda, "Enhancing the productivity of ultisols and oxisols in Malaysia using basalt and/or compost," Pedologist, vol. 55, no. 3, pp. 382–391, 2012.
2. J. P. Caliman, R. Carcasses, N. Perel, et al., "Agri-environmental indicators for sustainable palm oil production," Palmas, vol. 28, pp. 434–445, 2007.
3. K. J. Goh and R. Härdter, "General oil palm nutrition," in Oil Palm: Management for Large and Sustainable Yields, T. Fairhurst and R. Härdter, Eds., pp. 191–230, PPI/PPIC and IPI, Singapore, 2003.

4. Food and Agriculture Organization of the United Nations (FAO), Fertilizer Use by Crop in Malaysia, FAO, Rome, Italy, 2004.
5. I. Comte, F. Colin, J. K. Whalen, O. Grünberger, and J.-P. Caliman, "Agricultural practices in oil palm plantations and their impact on hydrological changes, nutrient fluxes and water quality in indonesia. A review," Advances in Agronomy, vol. 116, pp. 71–124, 2012.
6. S. Bricker, B. W. Longstaff, A. Dennison, K. Jones, C. Boicourt, and J. Woerner, "Effects of nutrient enrichment in the nations estuaries: a decade of change," in NOAA Coastal Ocean Program Decision Analysis, Series No. 26, National Centers for Coastal Ocean Science, Silver Spring, Md, USA, 2007.
7. A. M. Freeman, E. C. Lamon III, and C. A. Stow, "Nutrient criteria for lakes, ponds, and reservoirs: a Bayesian TREED model approach," Ecological Modelling, vol. 220, no. 5, pp. 630–639, 2009.
8. E. E. Aziz and S. M. El-Asry, "Efficiency of slow release Urea fertilizer on herb yield and essential oil production of lemon balm (Melissa officinalis L.) Plant," American-Eurasian Journal of Agricultural & Environmental Sciences, vol. 5, no. 2, pp. 141–147, 2009.
9. R. Prasad, "Efficient fertilizer use: the key to food security and better environment," Journal of Tropical Agriculture, vol. 47, no. 1-2, pp. 1–17, 2009.
10. P. P. Motavalli, K. W. Goyne, and R. P. Udawatta, "Environmental impacts of enhanced-efficiency nitrogen fertilizers," Crop Management, vol. 7, no. 1, 2008.
11. K. A. Nelson, P. C. Scharf, L. G. Bundy, and P. Tracy, "Agricultural management of enhanced-efficiency fertilizers in the north-central United States," Crop Management, vol. 7, no. 1, 2008.
12. A. Jarosiewicz and M. Tomaszewska, "Controlled-release NPK fertilizer encapsulated by polymeric membranes," Journal of Agricultural and Food Chemistry, vol. 51, no. 2, pp. 413–417, 2003.
13. A. D. Blaylock, J. Kaufmann, and R. D. Dowbenko, "Nitrogen fertilizer technologies," in Proceedings of the Western Nutrient Management Conference, vol. 6, Salt Lake City, Utah, USA, 2005.

14. International Fertilizer Industry Association (IFA), "Mineral Fertilizer Use and the Environment," 2000.

15. T. Kenzo, R. Yoneda, Y. Matsumoto, M. A. Azani, and N. M. Majid, "Leaf photosynthetic and growth responses on four tropical tree species to different light conditions in degraded tropical secondary forest, Peninsular Malaysia," Japan Agricultural Research Quarterly, vol. 42, no. 4, pp. 299–306, 2008.

16. G. W. Gee and J. W. Bauder, "Particle size analysis," in Methods of Soil Analysis. Part 1. Physican and Mineralogical Methods, A. Klute, Ed., pp. 383–411, ASA-SSSA, Madison, Wis, USA, 1986.

17. C. B. S. Teh and J. Talib, Soil Physics Analysis, vol. 1, Universiti Putra Malaysia Press, Serdang, Malaysia, 2006.

18. A. A. Ibitoye, Laboratory Manual on Basic Soil Analysis, Foladave, Akure, Nigeria, 2nd edition, 2006.

19. J. B. Jones Jr., Laboratory Guide for Conducting Soil Tests and Plant Analysis, pp. 81-82, CRC Press, Boca Ranton, Fla, USA, 2001.

20. A. Cottenie, Soil and Plant Testing as a Basis of Fertilizer Recommendations, FAO Soil Bulletin 38/2, FAO, Rome, Italy, 1980.

21. K. H. Tan, Soil Sampling, Preparation and Analysis, Taylor and Francis/CRC Press, Boca Raton, Fla, USA, 2nd edition, 2005.

22. G. W. Thomas, "Exchangeable cations," in Methods of Soil Analyes, A. L. Page, R. H. Miller, and D. R. Keeny, Eds., pp. 159–165, American Society of Agronomy, Madison, Wis, USA, 1982.

23. R. H. Bray and L. T. Kurtz, "Determination of total, organic and available forms of phosphorus in soils,"Soil Science, vol. 59, pp. 39–45, 1945.

24. S. Paramananthan, Soils of Malaysia: Their Characteristics and Identification, Academy of Sciences Malaysia, 2000.

25. PORIM, Environmental Impacts of Oil Palm Plantations in Malaysia, Palm Oil Research Institute of Malaysia, 1994.

26. A. E. Hartemink, "Soil erosion: perennial crop plantations," in Encyclopedia of Soil Science, Marcel Dekker, New York, NY, USA, 2006. · View at Google Scholar

27. R. H. V. Corley and P. B. Tinker, The Oil Palm, John Wiley & Sons/CRC Press, Hoboken, NJ, USA, 4th edition, 2003.

28. L. M. Maene, K. C. Tong, T. S. Ong, and A. M. Mokhtaruddin, "Surface wash under mature oil palm," inProceedings of the Symposium on Water in Malaysian Agriculture, pp. 203–216, MSSS, Kuala Lumpur, Malaysia, 1979.
29. K. J. Goh, C. B. Teo, P. S. Chew, and S. B. Chiu, "Fertilizer management in oil palm: agronomic principles and field practices," in Fertilizer Management for Oil Palm Plantations, vol. 44, pp. 20–21, ISP North-east Branch, Sandakan, Malaysia, 1999.
30. M. Banabas, M. A. Turner, D. R. Scotter, and P. N. Nelson, "Losses of nitrogen fertiliser under oil palm in Papua New Guinea: 1. Water balance, and nitrogen in soil solution and runoff," Australian Journal of Soil Research, vol. 46, no. 4, pp. 332–339, 2008.
31. K. J. Goh, R. Härdter, and T. Fairhurst, "Fertilizing for Maximum return," in Oil Palm: Management for Large and Sustainable Yields, T. Fairhurst and R. Härdter, Eds., pp. 279–306, PPI/PPIC and IPI, Singapore, 2003.
32. K. K. Kee and P. S. Chew, "A13: Nutrient Losses through Surface Runoff and Erosion- Implications for Improved Fertilizer Efficiency in mature Oil Palm," Applied Agricultural Research Sdn. Bhd., Locked Bag no. 212, 1996.
33. C. B. Wallace, M. G. Burton, S. G. Hefner, and T. A. DeWitt, "Effect of preceding rainfall on sediment, nutrients, and bacteria in runoff from biosolids and mineral fertilizer applied to a hayfield in a mountainous region," Agricultural Water Management, vol. 130, pp. 113–118, 2013.
34. F. M. Ziadat and A. Y. Taimeh, "Effect of rainfall intensity, slope, land use and antecedent soil moisture on soil erosion in an arid environment," Land Degradation & Development, vol. 24, no. 6, pp. 582–590, 2013.
35. J. F. Martínez-Murillo, E. Nadal-Romero, D. Regüés, A. Cerdà, and J. Poesen, "Soil erosion and hydrology of the western Mediterranean badlands throughout rainfall simulation experiments: a review,"Catena, vol. 106, pp. 101–112, 2013.
36. J. S. Kim, S. Y. Oh, and K. Y. Oh, "Nutrient runoff from a Korean rice paddy watershed during multiple storm events in the growing

season," Journal of Hydrology, vol. 327, no. 1-2, pp. 128–139, 2006.

37. S. I. Ahmed, R. P. Rudra, B. Gharabaghi, K. Mackenzie, and W. T. Dickinson, "Within-storm rainfall distribution effect on soil erosion rate," ISRN Soil Science, vol. 2012, Article ID 310927, 7 pages, 2012.
38. P. J. A. Kleinman, M. S. Srinivasan, C. J. Dell, J. P. Schmidt, A. N. Sharpley, and R. B. Bryant, "Role of rainfall intensity and hydrology in nutrient transport via surface runoff," Journal of Environmental Quality, vol. 35, no. 4, pp. 1248–1259, 2006.
39. T. Fu, J. P. Ni, C. F. Wei, and D. T. Xie, "Research on nutrient loss from terra gialla soil in Three Gorges Region under different rainfall intensity," Journal of Soil and Water Conservation, vol. 16, no. 2, pp. 33–35, 83, 2002.
40. X. Chen, S. Q. Jiang, K. Z. Zhang, and Z. P. Bian, "Law of phosphorus loss and its affecting factors in red soil slope land," Journal of Soil Erosion and Soil and Water Conservation, vol. 5, no. 3, pp. 38–41, 1999.
41. K. Jin, W. M. Cornelis, D. Gabriels et al., "Residue cover and rainfall intensity effects on runoff soil organic carbon losses," Catena, vol. 78, no. 1, pp. 81–86, 2009.
42. J. P. Caliman, B. Dubos, B. Tailliez, P. Robin, X. Bonneau, and I. de Barros, "Oil palm mineral nutrition management: current situation and prospects," in Proceedings of the 14th International Oil Palm Conference, p. 33, Cartagena de Indias, Columbia, September 2003.
43. Z. Fan, J. A. Moore, and D. L. Wenny, "Growth and nutrition of container-grown ponderosa pine seedlings with controlled-release fertilizer incorporated in the root plug," Annals of Forest Science, vol. 61, no. 2, pp. 117–124, 2004.

Chapter 2

# Effect of Pre plant Irrigation, Nitrogen Fertilizer Application Timing, and Phosphorus and Potassium Fertilization on Winter Wheat Grain Yield and Water Use Efficiency

Jacob T. Bushong, D. Brian Arnall, and William R. Raun

`Department of Plant and Soil Sciences, Oklahoma State University, 368 Agricultural Hall, Stillwater, OK 74078, USA

## ABSTRACT

Pre plant irrigation can impact fertilizer management in winter wheat. The objective of this study was to evaluate the main and interactive effects of pre plant irrigation, N fertilizer application timing, and different N, P, and K fertilizer treatments on grain yield and WUE.

Several significant two-way interactions and main effects of all three factors evaluated were observed over four growing seasons for grain yield and WUE. These effects could be described by differences in rainfall and soil moisture content among years. Overall, grain yield and WUE were optimized, if irrigation or adequate soil moisture were available prior to planting. For rain-fed treatments, the timing of N fertilizer application was not as important and could be applied before planting or top dressed without much difference in yield. The application of P fertilizer proved to be beneficial on average years but was not needed in years where above average soil moisture was present. There was no added benefit to applying K fertilizer. In conclusion, N and P fertilizer management practices may need to be altered yearly based on changes in soil moisture from irrigation and/or rainfall.

# INTRODUCTION

Winter wheat (Triticum aestivum L.) is cultivated on approximately three million hectares of the of the United States' Central Rolling Red Plains, present in parts of Kansas, Oklahoma, and Texas [1]. The vast majority of these wheat hectares are cultivated under rain-fed conditions without irrigation. There are small isolated areas where wheat is grown with the aid of irrigation; one area in particular is the Lugert-Altus Irrigation District in southwestern Oklahoma. The irrigation water in this district is delivered to producer fields via canals from the Lake Lugert-Altus reservoir and applied through furrow or flood irrigation techniques. Though most of the water is utilized to irrigate cotton (Gossypium hirsutum L.), some producers have taken advantage of the available reservoir water in summer months to soften heavy textured surface soils in order to cultivate and prepare ground prior to wheat planting. The yearly amount of wheat hectares receiving irrigation in the counties that include the district ranges from zero hectares in years where there is not sufficient water in the reservoir to about 4,000 hectares [2].

Studies have shown that wheat planted into moist soil typically has increased emergence, stand establishment, and root growth, which can all lead to potentially higher grain yields [3]. The improved root development beneficially increases the potential soil water and nutrient reservoirs of the growing crop [4]. Researchers have reported that early season root growth can be stimulated with fertilizer, mainly nitrogen (N), which increases the potential for greater water extraction from the soil profile. Studies in wheat have reported that increases in N fertilizer rate, typically increase water use efficiency (WUE) in both irrigated systems [5–7] and rain-fed systems [8–10].

As previously stated, early increased root growth from adequate soil moisture can be advantageous for soil nutrient acquisition. This would be beneficial for nutrients that are immobile in the soil, such as P and K. Relationships between soil water dependent root growth and plant P uptake have been observed in cereal grains, mostly because of the effects of soil moisture on the movement of phosphorus via diffusion and root development [11, 12]. Researchers have observed that plants take up more native soil P in more moist soil environments and that the uptake of fertilizer P is not as sensitive to changes to soil moisture content [12–14]. A similar relationship between the plant availability of K and soil water dependent root growth also exists. Adequate or increased soil moisture content typically leads to increased root growth as well as the diffusive flux of K to the root surface [15–17]. Providing adequate soil moisture content has also been shown to increase the efficiency of K fertilizer applications [17, 18].

The interactive effects of irrigation and N, P, and K fertilization have been evaluated with N most commonly being evaluated with P or K or the three together. The response to N fertilization is typically almost always observed regardless of irrigation or soil moisture content, but a response to P and/or K fertilization along with N fertilization is dependent on the amount of soil moisture, the timing the moisture is received by the growing crop, and the soil type [19–21].

Because N is mobile in the soil and taken up by the plant via mass flow mechanisms, N is typically taken up in greatest quantities during periods of active growth [22]. Much research has been conducted comparing yields and N fertilizer recovery of application timings and amounts in winter wheat. Some research has reported little to no added benefit to grain yield from spring or split N fertilization applications [23], but others have reported that significant grain yield increases when fertilization is split or spring applied [24, 25]. One consensus that has been reached is that split or spring N fertilizer applications increases the recovery of fertilizer N in the grain [23, 24, 26, 27].

The objective of this paper was to evaluate the effect of preplant irrigation, N fertilizer application timing, and P and K fertilization on winter wheat grain yields and WUE on a long-term soil fertility experiment site. The results of this evaluation will be used to assist in making proper N, P, and K fertilizer recommendations for optimizing grain yield and WUE in in the Central Rolling Red Plains.

# MATERIALS AND METHODS

This experiment was conducted at the Oklahoma State University Southwest Research and Extension Center located near Altus, Oklahoma. The soil type for the study area is a Hollister (Fine, smectitic, thermic Typic Haplusterts) silty clay loam [28]. The Hollister soil series is mapped extensively on over 900,000 hectares in the Central Rolling Red Plains and is used mainly for crop production [29]. The data was collected from a long-term winter wheat N, P, and K fertilizer trial that was established in 1966.

Treatments for the study area were split into areas that received either a single preplant irrigation or were strictly rain-fed. The N fertilizer treatments were applied all at once prior to planting or midseason just prior to first hollow stem at a rate of 45 kg N ha$^{-1}$ or 90 kg N ha$^{-1}$ as urea (46-0-0). The P fertilizer treatments were applied all at once prior to planting at a rate of 20 kg P ha$^{-1}$ as triple super phosphate (0-20-0). The K fertilizer treatments were also applied all at once prior to planting at a rate 37 kg K ha$^{-1}$ as muriate of potash (0-0-50). A detailed list, with an assigned fertilizer treatment number, of the combination N-P-K fertilizer treatments along with an unfertilized check are described in Table 1. Irrigation water was always applied in late July or the first part of August using flood irrigation techniques at a rate of 100 mm. Fertilizer treatments that were applied prior to planting were broadcast applied 30 to 45 days after irrigation and 45 to 60 days prior to planting and incorporated using conventional tillage techniques. Plots were broadcast seeded at a rate of approximately 100 kg ha$^{-1}$. Planting took place around the first of October and grain harvest occurred around the first of June. Best agronomic practices were employed for pest management and control. Specific dates of agronomic activities for each site year analyzed are reported in Table 2.

**Table 1**: Fertilizer rates and average soil test P and K for treatments in this study. N fertilizer was applied as urea (46-0-0) either pre plant or midseason depending upon subplot assignment. P and K fertilizer treatments were all applied pre plant as triple super phosphate (0-20-0) and muriate of potash (0-0-50), respectively

| Treatment | | | | Soil test values [a] | | | |
|---|---|---|---|---|---|---|---|
| | Fertilizer treatment | | | Irrigated | | Rain-fed | |
| | kg N ha$^{-1}$ | **kg P ha$^{-1}$** | **kg K ha$^{-1}$** | **mg P kg$^{-1}$** | **mg K kg$^{-1}$** | **mg P kg$^{-1}$** | **mg K kg$^{-1}$** |
| 1 | 0 | 0 | 0 | 6 | 302 | 4 | 334 |
| 2 | 45 | 0 | 0 | 6 | 319 | 6 | 350 |
| 3 | 90 | 0 | 0 | 6 | 320 | 7 | 350 |
| 4 | 45 | 20 | 0 | 21 | 322 | 19 | 337 |
| 5 | 90 | 20 | 0 | 19 | 312 | 27 | 337 |
| 6 | 45 | 20 | 37 | 19 | 328 | 19 | 368 |
| 7 | 90 | 20 | 37 | 17 | 320 | 19 | 373 |

[a] Soil test p and k values derived from Mehlich 3 solution extraction (Mehlich, 1984 [30])

**Table 2**: Dates of agronomic activities for growing seasons utilized in this study

| **Year** | **Irrigation [a]** | **Pre plant application** | **Planting** | **Top dress application** | **Harvest** |
|---|---|---|---|---|---|
| 2003 | Aug. 10, 2002 | Sept. 10, 2002 | Oct. 1, 2002 | Mar. 7, 2003 | May 30, 2003 |
| 2004 | Aug. 5, 2003 | Sept. 17, 2003 | Oct. 1, 2003 | Mar. 11, 2004 | May 28, 2004 |
| 2008 | Aug. 8, 2007 | Aug. 29, 2007 | Oct. 2, 2007 | Feb. 28, 2008 | Jun. 3, 2008 |
| 2011 | Jul. 27, 2010 | Sept. 13, 2010 | Oct. 1, 2010 | Mar. 15, 2011 | Jun. 1, 2011 |

[a] Irrigation applied by flood at a rate of 100 mm

Plots for the irrigated and rain-fed treatments were 5.7 by 18.3 meters and 8.5 by 30.5 meters, respectively. Individual plots were harvested with a self-propelled, small plot grain combine and grain yields were adjusted to 12.5 percent moisture. Water use efficiency (WUE) was calculated for each treatment as the grain yield per unit of area per amount of water added through irrigation and rainfall and reported as kg ha$^{-1}$ mm$^{-1}$. Weather data, which includes daily precipitation, temperature, and the 0 to 40 cm fractional water index (FWI), was downloaded from the nearby Oklahoma Mesonet [31] climate monitoring station. The FWI is a normalized value which ranges from 0.00 for very dry soil to 1.00 for soil at field capacity [32]. A summary of the collected weather data is displayed in Figures 1 and 2.

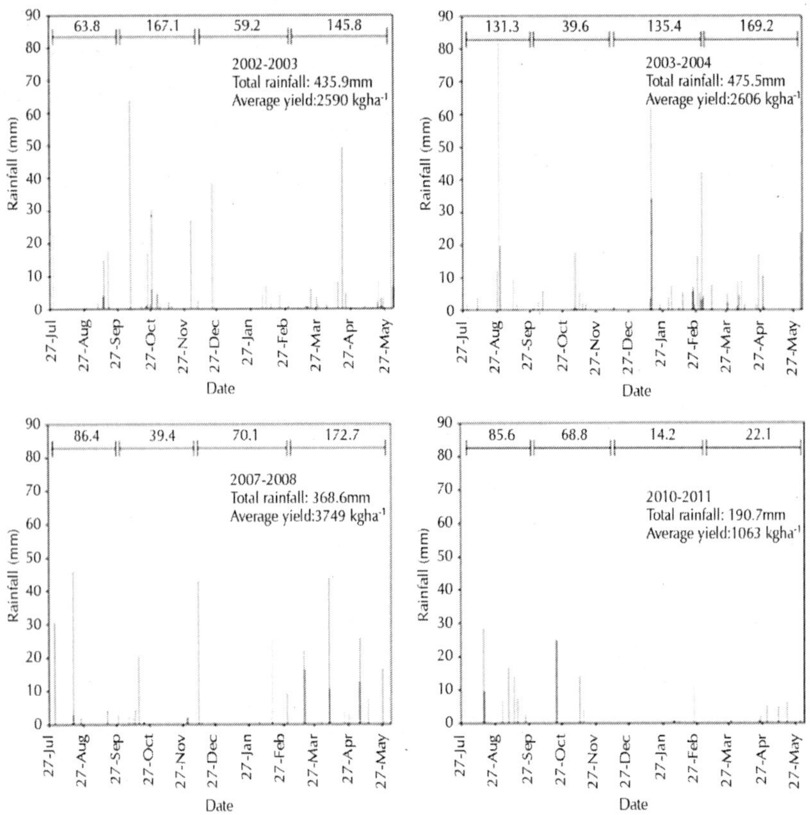


**Figure 1**: Rainfall distribution for Altus, Oklahoma, area from the average time of pre plant irrigation to the average time of grain harvest for years analyzed. Data obtained from the Oklahoma Mesonet climate monitoring station [31].

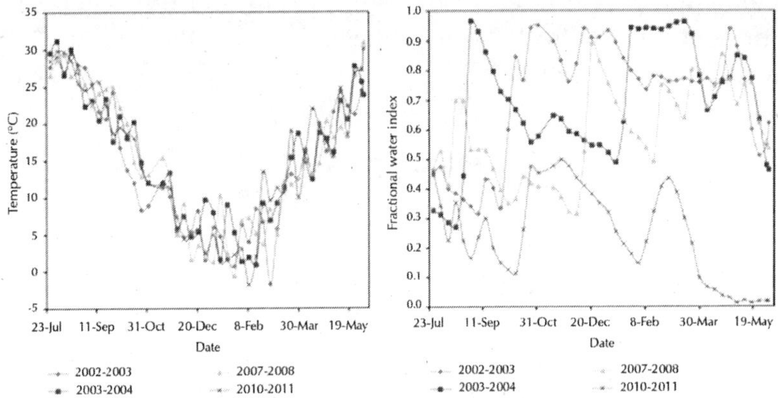


**Figure 2:** Average weekly air temperature and fractional water index for Altus, Oklahoma, from the average time of pre plant irrigation to the average time of grain harvest for years analyzed. Data obtained from the Oklahoma Mesonet climate monitoring station [31].

Growing seasons to be analyzed for this trial were selected by years that contained accurate irrigation data and reliable weather data and years where the wheat crop was taken to grain harvest. Prior to the 2001-2002 growing season, all N fertilizer treatments were applied pre plant. After the 2001-2002 growing season, the N fertilizer treatments were split where half the N fertilizer treatments were applied all pre plant or all midseason. After all the selection criteria were applied, four site years (2002-2003, 2003-2004, 2007-2008, and 2010-2011) were chosen for analysis. From this point forward, the site years will be referred to by the year of their grain harvest.

Soil test P and soil test K values had been collected at random times throughout the duration of this trial and were derived from a Mehlich 3 solution extraction [30]. Typically, soil samples were collected at a depth from 0 to 15 cm prior to fertilization and planting in the late summer. An average soil test value from over the entirety of this trial for each treatment is provided in Table 1.

The experimental design was a split-split plot design with three replications. The main plot was water (irrigated, rain-fed). The subplot was N fertilizer application timing (pre plant and midseason). The sub-subplot was the N, P, and/or K fertilizer rate treatment (Table 1). Analysis of variance was used to determine significant (alpha = 0.10) main and interaction effects of treatments on grain yield and WUE. Non orthogonal contrasts were utilized to determine differences in specific treatment groupings.

# RESULTS

## Effect on Grain Yield

Analysis of variance showed that there is no significant three-way interaction between irrigation, N fertilizer application time, and fertilizer treatment for any of the four growing seasons (Table3). However, all three two-way interactions were significant for the 2003, 2008, and 2011 growing seasons. In 2004, there were no significant interaction effects, but all three main effects were observed to be significant (Table 3).

**Table 3**: Analysis of variance for main effects and interaction effects of factors affecting grain yield (GY) and water use efficiency (WUE) utilized in this study by growing season

| Source | Year | | | | | | | |
|---|---|---|---|---|---|---|---|---|
| | **2003** | | **2004** | | **2008** | | **2011** | |
| | **GY** | **WUE** | **GY** | **WUE** | **GY** | **WUE** | **GY** | **WUE** |
| Water | * | ** | * | ** | * | * | ** | ns |
| N timing | ns | ns | * | * | ns | ns | ns | ns |
| N timing × water | * | * | ns | ns | * | * | * | ns |
| Treatment | *** | *** | *** | *** | *** | *** | *** | *** |
| Treatment × N timing | * | * | ns | * | * | * | * | * |
| Treatment × water | *** | ** | ns | ns | * | ns | *** | ** |
| Treatment × N timing × water | ns | ns | ns | ns | ns | ns | ns | ns |

*, **, and *** are significant at the 0.1, 0.01, and 0.001 level, respectively.

ns: not significant at the 0.1 level.

For the two-way interaction of irrigation and N fertilizer application timing for 2003, 2008, and 2011, grain yields were typically higher for treatments that were irrigated prior to planting (Table4). Grain yields increased from 134 to 456 kg ha$^{-1}$ for irrigated plots that received N fertilizer pre plant compared to irrigated plots that received top dress N fertilizer in February or March. Minimal difference was observed between rain-fed treatments that received N pre plant compared to those that received top dress N. Rain-fed treatments that received top dress N fertilizer applications in the spring had from 51 to 285 kg ha$^{-1}$ higher grain yields than rain-fed treatments that received N fertilizer prior to planting.

**Table 4**: Grain yield and water use efficiency (WUE) interaction means by year for plots that received either irrigation or were rain-fed and received their total N fertilizer pre plant or top dressed midseason

| | Year | | | Year | |
|---|---|---|---|---|---|
| | 2003 | 2008 | 2011 | 2003 | 2008 |
| | Grain yield (kg ha$^{-1}$) | | | WUE (kg ha$^{-1}$ mm$^{-1}$) | |
| Irrigated | | | | | |
| Pre plant | 2730 | 4213 | 1397 | 5.1 | 9.0 |
| Top dress | 2596 | 3757 | 1106 | 4.8 | 8.0 |
| Rain-fed | | | | | |
| Pre plant | 2438 | 3355 | 849 | 5.6 | 9.1 |
| Top dress | 2598 | 3640 | 900 | 6.0 | 9.9 |
| SED [a] | 56 | 161 | 65 | 0.1 | 0.4 |
| | | | | | |

[a] Standard error of the difference for the interaction of water and N application timing

When comparing the interaction of N fertilizer application timing and fertilizer treatment, grain yields were typically higher in plots that received 90 kg N ha$^{-1}$ prior to planting and had P fertilizer added in both 2003 and 2008 (Table 5). Single degree-of-freedom contrasts revealed a few significant differences (Table 5). Treatments that received 45 kg N ha$^{-1}$ pre plant were not significantly different from plots that received 45 kg N ha$^{-1}$ top dress for 2008 and 2011; however, treatments in 2003 that received 45 kg N ha$^{-1}$ top dress were significantly higher than those that received the same amount of N prior to planting. Treatments that received 90 kg N ha$^{-1}$ pre plant had significant increases in grain yield by 247 kg ha$^{-1}$ and 338 kg ha$^{-1}$ for 2003 and 2008, respectively, but no significant difference was detected for 2011. The addition of P and K was not affected by N fertilizer timing in 2003 and 2011, but significant grain yield increases were observed for plots that received both P and K fertilizer and received N prior to planting instead of top dress in the spring in 2008. When investigating the response to P and K fertilization for N fertilizer application times separately, yield increases were observed for each N application time and addition of P fertilizer for both 2003 and 2008. However, significant yield decreases were observed for both N application times with the addition of P in 2011. The addition of K fertilizer paired with pre plant N application significantly decreased yields for both 2003 and 2011 and did not have any effect on yield in 2008. No significant differences in K fertilizer responses were observed when N fertilizer was top dress applied in the spring of 2003, 2008, and 2011.

**Table 5**: Grain yield means and selected contrasts by year for plots with different fertilizer treatments and those received either their total N fertilizer pre plant or top dressed midseason

(a)

| | **Fertilizer treatment** | **Year** | | |
|---|---|---|---|---|
| | | **2003** | **2008** | **2011** |
| | | **Grain yield (kg ha$^{-1}$)** | | |
| Pre plant N | 1 | 1575 | 2582 | 1055 |
| | 2 | 2332 | 3593 | 1342 |
| | 3 | 2671 | 3669 | 1252 |
| | 4 | 2749 | 4058 | 1099 |
| | 5 | 3274 | 4238 | 1129 |
| | 6 | 2498 | 3849 | 1125 |
| | 7 | 2989 | 4498 | 857 |

| | | | | |
|---|---|---|---|---|
| Top dress N | 1 | 1688 | 2902 | 964 |
| | 2 | 2584 | 3848 | 1103 |
| | 3 | 2435 | 3753 | 1133 |
| | 4 | 2955 | 3863 | 1069 |
| | 5 | 2882 | 4106 | 840 |
| | 6 | 2753 | 3885 | 1015 |
| | 7 | 2877 | 3533 | 898 |
| | SED [a] | 176 | 259 | 91 |

(b)

| Contrasts | Year | | | | | |
|---|---|---|---|---|---|---|
| | 2003 | | 2008 | | 2011 | |
| | Sig | Dif | **Sig** | **Dif** | **Sig** | **Dif** |
| 45 kg N ha$^{-1}$ at pre plant versus 45 kg N ha$^{-1}$ at top dress | * | −238 | ns | −32 | ns | 127 |
| 90 kg N ha$^{-1}$ at pre plant versus 90 kg N ha$^{-1}$ at top dress | * | 247 | * | 338 | ns | 122 |
| 45 kg N ha$^{-1}$ versus 90 kg N ha$^{-1}$ | ** | −210 | ns | −117 | *** | 107 |
| Pre plant 45 kg N ha$^{-1}$ versus 90 kg N ha$^{-1}$ | *** | −452 | * | −301 | * | 109 |
| Top dress 45 kg N ha$^{-1}$ versus 90 kg N ha$^{-1}$ | ns | 33 | ns | 68 | * | 105 |
| P fertilizer with pre plant N versus with top dress N | ns | 10 | * | 314 | ns | 97 |
| K fertilizer with pre plant N versus with top dress N | ns | −72 | * | 465 | ns | 35 |
| Pre plant N-P fertilizer added versus no P fertilizer added | *** | 510 | ** | 517 | ** | −183 |
| Top dress N-P fertilizer added versus no P fertilizer added | ** | 409 | ns | 183 | ** | −163 |
| Pre plant N-K fertilizer added versus no K fertilizer added | * | −268 | ns | 26 | * | −123 |
| Top dress N-K fertilizer added versus no K fertilizer added | ns | −104 | ns | −275 | ns | 2 |

[a] Standard error of the difference for the interaction of N application timing and fertilizer treatment

*, **, and *** are significant at the 0.1, 0.01, and 0.001 level, respectively

ns: not significant at the 0.1 level.

For the interaction effect of irrigation and fertilizer treatment on grain yield, treatments that received irrigation prior to planting had higher grain yields in 2003, 2008, and 2011 (Table 6). Single degree-of-freedom contrasts revealed numerous significant differences (Table

6). Irrigated plots that received both rates of N and P fertilizer and K fertilizer had increased yields compared to the same rain-fed treatments. In 2003 and 2008, regardless of N application time, the 90 kg N ha$^{-1}$ rate of N fertilizer treatment had higher grain yields than the 45 kg N ha$^{-1}$ rate. The same trend was observed for the irrigated treatment in 2003 and 2008; however, in the rain-fed treatments no significant differences were observed in grain yields between the two N fertilizer rates for 2003 and 2008. In 2011, the differences between N fertilizer rates were significantly higher for the 45 kg N ha$^{-1}$ rate. This was true regardless of irrigation treatment and when irrigated and/ or rain-fed sites were analyzed separately. The addition of P fertilizer significantly increased yields in irrigated treatments in 2003 and 2008 but significantly decreased yields in 2011. No differences were detected in 2003 and 2008 with the addition of P fertilizer in rain-fed plots, but, like the irrigated treatments, the addition of P fertilizer significantly decreased yields in 2011. The addition of K fertilizer appeared to have little effect on grain yield, regardless whether the plots were irrigated or rain-fed. The only significant difference occurred in 2003 when addition of K decreased yields in irrigated treatments. When grouping the fertilizer treatments that received 90 kg N ha$^{-1}$ and P fertilizer, the addition of P fertilizer in irrigated treatments increased yields in 2003 and 2008, but decreased yield in 2011. For the other treatments that received 90 kg N ha$^{-1}$, there was no difference in P fertilization under rain-fed conditions and no meaningful differences were observed with the addition of K fertilizer under irrigated and rain-fed conditions.

**Table 6**: Grain yield means and selected contrasts by year for plots with different fertilizer treatments and those either received a flush of irrigation prior to planting or did not receive irrigation.

(a)

| | | | **Year** | |
|---|---|---|---|---|
| | **Fertilizer treatment** | **2003** | **2008** | **2011** |
| | | | **Grain yield (kg ha$^{-1}$)** | |
| | 1 | 1452 | 2726 | 1097 |
| | 2 | 2508 | 3900 | 1369 |
| | 3 | 2465 | 3887 | 1512 |

| Irrigated | 4 | 3126 | 4112 | 1344 |
|---|---|---|---|---|
| | 5 | 3393 | 4623 | 1129 |
| | 6 | 2541 | 4158 | 1229 |
| | 7 | 3152 | 4488 | 1079 |
| | 1 | 1811 | 2757 | 923 |
| | 2 | 2408 | 3541 | 1076 |
| | 3 | 2641 | 3536 | 873 |
| Rain-fed | 4 | 2579 | 3809 | 823 |
| | 5 | 2763 | 3720 | 840 |
| | 6 | 2710 | 3576 | 911 |
| | 7 | 2713 | 3542 | 677 |
| | SED [a] | 169 | 235 | 72 |

**(b)**

| Contrasts | Year | | | | | |
|---|---|---|---|---|---|---|
| | 2003 | | **2008** | | **2011** | |
| | Sig | Dif | Sig | Dif | Sig | Dif |
| P fertilizer added irrigated versus rain-fed | *** | 362 | *** | 684 | *** | 382 |
| K fertilizer added irrigated versus rain-fed | ns | 135 | *** | 764 | *** | 360 |
| 45 kg N $ha^{-1}$ irrigated versus 45 kg N $ha^{-1}$ rain-fed | * | 159 | ** | 415 | *** | 377 |
| 90 kg N $ha^{-1}$ irrigated versus 90 kg N $ha^{-1}$ rain-fed | *** | 298 | *** | 734 | *** | 443 |
| 45 kg N $ha^{-1}$ versus 90 kg N $ha^{-1}$ | ** | −210 | ns | −117 | *** | 107 |
| Irrigated 45 kg N $ha^{-1}$ versus 90 kg N $ha^{-1}$ | * | −279 | * | −276 | * | 74 |
| Rain-fed 45 kg N $ha^{-1}$ versus 90 kg N $ha^{-1}$ | ns | −140 | ns | 43 | ** | 140 |
| Irrigated P fertilizer added versus no P fertilizer added | *** | 772 | ** | 474 | *** | −204 |
| Rain-fed P fertilizer added versus no P fertilizer added | ns | 146 | ns | 226 | ** | −142 |
| Irrigated K fertilizer added versus no K fertilizer added | ** | −413 | ns | −44 | ns | −83 |
| Rain-fed K fertilizer added versus no K fertilizer added | ns | 41 | ns | −206 | ns | 39 |
| Irrigated, 90 kg N $ha^{-1}$, P fertilizer added versus no P fertilizer added | *** | 928 | ** | 736 | *** | −383 |

| | | | | | | |
|---|---|---|---|---|---|---|
| Rain-fed, 90 kg N $ha^{-1}$, P fertilizer added versus no P fertilizer added | ns | 121 | ns | 184 | ns | −33 |
| Irrigated, 90 kg N $ha^{-1}$, K fertilizer added versus no K fertilizer added | ns | −241 | ns | −135 | ns | −50 |
| Rain-fed, 90 kg N $ha^{-1}$, K fertilizer added versus no K fertilizer added | ns | −49 | ns | 178 | * | −164 |

[a] Standard error of the difference for the interaction of fertilizer treatment and water treatment.

*, **, and *** are significant at the 0.1, 0.01, and 0.001 level, respectively.

ns: not significant at the 0.1 level.

In 2004, significant differences among the main effects of fertilizer treatment and N application time were observed. Overall, treatments that received 90 kg N $ha^{-1}$ compared to 45 kg N $ha^{-1}$ had higher grain yields (Table 7). Single degree-of-freedom contrasts showed that there was no effect of adding P or K fertilizer in 2004 (Table 7). Grain yields of treatments receiving pre plant N fertilizer applications were significantly higher than top dress N fertilized treatments, but only by 139 kg $ha^{-1}$.

**Table 7:** Grain yield means and selected contrasts for the significant main effects of fertilizer treatment and N application timing for 2004 growing season

(a)

| Fertilizer treatment | Grain yield (kg $ha^{-1}$) | N application time | Grain yield **(kg $ha^{-1}$)** |
|---|---|---|---|
| 1 | 1569 | Pre plant | 2676 |
| 2 | 2688 | Top dress | 2537 |
| 3 | 2852 | **SED [a]** | **61** |
| 4 | 2607 | | |
| 5 | 2921 | | |
| 6 | 2725 | | |
| 7 | 2883 | | |
| **SED [a]** | **151** | | |

**(b)**

| Contrasts | Sig | Dif |
|---|---|---|
| P fertilizer added versus No P fertilizer added | ns | −6 |
| K fertilizer added versus No K fertilizer added | ns | 40 |
| 45 kg N ha$^{-1}$ versus 90 kg N ha$^{-1}$ | | -212 |

[a] SED: standard error of the difference between the main effects of fertilizer treatment and N application timing.

* is significant at the 0.1 level.

ns: not significant at the 0.1 level.

## Effect on Water Use Efficiency

Analysis of variance on WUE found no significant three-way interaction for irrigation, N application time, and fertilizer treatment for site-years (Table 3). The 2003 and 2011 growing seasons had a significant two-way interaction for irrigation and fertilizer treatment. All four site-years had a significant two-way interaction for N fertilizer application timing and fertilizer treatment (Table 3). The interaction of irrigation and N fertilizer application timing was only significant for 2003 and 2008. It should be noted that irrigation had a significant effect for 2003, 2004, and 2008; however, drought conditions were so severe in 2011 that no differences were observed.

In 2003 and 2008, the highest WUE for the interactive effect of irrigation and timing of N fertilizer application was reported for the rain-fed treatments that received top dress N in the spring (Table 4). Both N application timings in the rain-fed treatments had higher WUE values compared to the irrigated treatments in both 2003 and 2008. When analyzing only the irrigated treatments, those that received pre plant N fertilizer had significantly higher WUE values for both years (Table 4). Though the trends among treatments were the same for both years, the magnitudes of WUE were different for each individual year. The treatments in 2008 yielded around 40 percent higher WUE values compared to 2003.

As previously stated, the two-way interaction of N fertilizer application time and fertilizer treatment on WUE was significant for all four years. The 2008 growing season had the highest WUE values across all treatments, whereas the 2011 growing season had the lowest WUE values across all treatments. This is likely because 2008 had the highest grain yield for all years evaluated and 2011 had the lowest grain yields for all years evaluated. Fertilizer treatment groupings that were partitioned using single degree-of-freedom contrasts revealed several significant differences (Table 8). The 45 kg N $ha^{-1}$ top dress application increased WUE for each year except for 2011. However, the pre plant 90 kg N $ha^{-1}$ application compared to top dress increased WUE for all four years. The treatments receiving 90 kg N $ha^{-1}$ compared to the lower N rates displayed higher WUE values for 2003, 2004, and 2008, especially those treatments that were irrigated. However, in 2011, the 45 kg N $ha^{-1}$ treatments had significantly higher WUE values than the higher N rates for both the irrigated and rain-fed treatments. Treatments that received P fertilizer and pre plant N fertilizer had increases in WUE for 2004, 2008, and 2011 but decreased WUE in 2003. There were no significant effects of N application timing on K fertilized plots for 2003, 2004, and 2011, but WUE was significantly higher in 2008 for K fertilized plots that received N prior to planting compared to a top dress N application. For plots that received N fertilizer prior to planting the addition of P increased WUE in 2003, 2004, and 2008 compared to plots that did not receive any P fertilizer. The opposite was true in 2011, where the addition of P fertilizer decreased WUE on treatments that received N prior to planting. Treatments that received top dress N fertilizer applications in the spring were not significantly affected by the addition of P fertilizer in 2004 and 2008, but an increase in WUE was observed in 2003 and a decrease in WUE was observed in 2011. The addition of K fertilizer in treatments that received pre plant N fertilizer applications compared to treatments that did not receive K fertilizer were not significantly different in 2004 and 2008, but WUE was significantly decreased in 2003 and 2011. No significant decreases or increases in WUE were observed in treatments that received top dress N applications and did or did not receive K fertilization.

**Table 8**: Water use efficiency (WUE) means and selected contrasts by year for plots with different fertilizer treatments and those received either their total N fertilizer pre plant or top dressed midseason

(a)

| | Fertilizer treatment | Year | | | |
|---|---|---|---|---|---|
| | | 2003 | 2004 | 2008 | 2011 |
| | | WUE (kg ha$^{-1}$ mm$^{-1}$) | | | |
| Pre plant N | 1 | 3.3 | 3.1 | 6.2 | 4.5 |
| | 2 | 4.8 | 5.0 | 8.6 | 5.7 |
| | 3 | 5.5 | 5.5 | 8.7 | 5.0 |
| | 4 | 5.6 | 4.9 | 9.7 | 4.4 |
| | 5 | 6.7 | 6.2 | 10.1 | 4.7 |
| | 6 | 5.2 | 5.2 | 9.2 | 4.6 |
| | 7 | 6.2 | 5.9 | 10.7 | 3.4 |
| Top dress N | 1 | 3.5 | 2.9 | 7.1 | 4.1 |
| | 2 | 5.4 | 5.3 | 9.3 | 4.7 |
| | 3 | 5.1 | 5.4 | 9.1 | 4.7 |
| | 4 | 6.1 | 5.1 | 9.3 | 4.5 |
| | 5 | 5.9 | 4.9 | 9.9 | 3.6 |
| | 6 | 5.8 | 5.3 | 9.4 | 4.4 |
| | 7 | 5.9 | 5.2 | 8.4 | 3.8 |
| SED [a] | SED | 0.3 | 0.4 | 0.7 | 0.3 |

(b)

| Contrasts | Year | | | | | | | |
|---|---|---|---|---|---|---|---|---|
| | 2003 | | 2004 | | **2008** | | **2011** | |
| | **Sig** | **Dif** | **Sig** | **Dif** | **Sig** | **Dif** | **Sig** | **Dif** |
| 45 kg N ha$^{-1}$ at pre plant versus 45 kg N ha$^{-1}$ at top dress | ** | −0.6 | ns | −0.2 | ns | −0.2 | ns | 0.4 |
| 90 kg N ha$^{-1}$ at pre plant versus 90 kg N ha$^{-1}$ at top dress | * | 0.5 | ** | 0.7 | ns | 0.7 | ns | 0.3 |
| 45 kg N ha$^{-1}$ versus 90 kg N ha$^{-1}$ | ** | −0.4 | * | −0.4 | ns | −0.2 | *** | 0.5 |
| Pre plant 45 kg N ha$^{-1}$ versus 90 kg N ha$^{-1}$ | *** | −1.0 | *** | −0.8 | * | −0.7 | ** | 0.5 |
| Top dress 45 kg N ha$^{-1}$ versus 90 kg N ha$^{-1}$ | ns | 0.1 | ns | 0.1 | ns | 0.2 | * | 0.5 |

| | | | | | | | | |
|---|---|---|---|---|---|---|---|---|
| P fertilizer with pre plant N versus with top dress N | ns | −0.1 | * | 0.4 | ns | 0.6 | ns | 0.2 |
| K fertilizer with pre plant N versus with top dress N | ns | −0.2 | ns | 0.3 | * | 1.0 | ns | −0.1 |
| Pre plant N-P fertilizer added versus no P fertilizer added | *** | 1.0 | ns | 0.3 | ** | 1.2 | ** | −0.8 |
| Top dress N-P fertilizer added versus no P fertilizer added | ** | 0.7 | ns | −0.4 | ns | 0.4 | ** | −0.7 |
| Pre plant N-K fertilizer added versus no K fertilizer added | * | −0.5 | ns | −0.1 | ns | 0.1 | * | −0.5 |
| Top dress N-K fertilizer added versus no K fertilizer added | ns | −0.2 | ns | 0.2 | ns | −0.7 | ns | 0.1 |

[a]Standard error of the difference for the interaction of N application timing and fertilizer treatment.

* and ** are significant at the 0.1 level and 0.01 level, respectively.

ns: not significant at the 0.1 level.

The only years to have the significant two-way interaction of irrigation and fertilizer treatment were 2003 and 2011. In 2003, irrigated plots displayed lower WUE values than rain-fed plots. In 2011, irrigated treatments had slightly lower WUE values but were not as high as 2003 (Table 9). Several groupings based upon fertilizer treatment and irrigated versus rain-fed moisture conditions exhibited significant differences based upon single degree-of-freedom contrasts (Table9). Irrigated treatments that received P or K fertilizer applications had decreased WUE for both years compared to rain-fed treatments. Treatments that received 45 kg N $ha^{-1}$ had lower WUE values in irrigated plots compared to rain-fed plots for both 2003 and 2011. The same trend was observed for plots that received 90 kg N $ha^{-1}$ in 2003, but there was no significant difference in WUE for irrigated versus rain-fed plots in 2011. For the 2003 growing season, the addition of P or K fertilizer compared to plots that did not receive P or K fertilizer revealed only two significant differences in WUE values. Irrigated treatments that did not receive K fertilizer had higher WUE values than treatments that received K fertilizer. Irrigated treatments that received 90 kg N $ha^{-1}$and P fertilizer application had increased WUE values compared to treatments that did not receive P fertilizer. In 2011, regardless whether the treatment was irrigated or rain-fed or whether the higher rate of N was applied, WUE values were always lower in plots that received P or K fertilizer.

**Table 9**: Water use efficiency (WUE) means and selected contrasts by year for plots with different fertilizer treatments and those either received a flush of irrigation prior to planting or did not receive irrigation

(a)

| | Fertilizer treatment | Year | |
|---|---|---|---|
| | | 2003 | 2011 |
| | | WUE (kg ha$^{-1}$ mm$^{-1}$) | |
| Irrigated | 1 | 2.7 | 3.8 |
| | 2 | 4.7 | 4.7 |
| | 3 | 4.6 | 5.2 |
| | 4 | 5.8 | 4.6 |
| | 5 | 6.3 | 3.9 |
| | 6 | 4.7 | 4.2 |
| | 7 | 5.9 | 3.7 |
| Rain-fed | 1 | 4.2 | 4.8 |
| | 2 | 5.5 | 5.6 |
| | 3 | 6.1 | 4.6 |
| | 4 | 5.9 | 4.3 |
| | 5 | 6.3 | 4.4 |
| | 6 | 6.2 | 4.8 |
| | 7 | 6.2 | 3.5 |
| | SED [a] | 0.3 | 0.3 |

(b)

| Contrasts | Year | | | |
|---|---|---|---|---|
| | **2003** | | **2011** | |
| | **Sig** | **Dif** | **Sig** | **Dif** |
| P fertilizer added irrigated versus rain-fed | *** | −0.5 | ns | −0.1 |
| K fertilizer added irrigated versus rain-fed | *** | −0.9 | ns | −0.2 |
| 45 kg N ha$^{-1}$ irrigated versus 45 kg N ha$^{-1}$ rain-fed | *** | −0.8 | * | −0.4 |
| 90 kg N ha$^{-1}$ irrigated versus 90 kg N ha$^{-1}$ rain-fed | *** | −0.6 | ns | 0.1 |
| 45 kg N ha$^{-1}$ versus 90 kg N ha$^{-1}$ | ** | −0.4 | *** | 0.5 |
| Irrigated 45 kg N ha$^{-1}$ versus 90 kg N ha$^{-1}$ | * | −0.5 | ns | 0.3 |
| Rain-fed 45 kg N ha$^{-1}$ versus 90 kg N ha$^{-1}$ | ns | −0.3 | *** | 0.7 |

| | | | | |
|---|---|---|---|---|
| Irrigated P fertilizer added versus no P fertilizer added | *** | 1.4 | ** | −0.7 |
| Rain-fed P fertilizer added versus no P fertilizer added | ns | 0.3 | ** | −0.7 |
| Irrigated K fertilizer added versus no K fertilizer added | ** | −0.8 | ns | −0.3 |
| Rain-fed K fertilizer added versus no K fertilizer added | ns | 0.1 | ns | −0.2 |
| Irrigated, 90 kg N $ha^{-1}$ and P fertilizer added versus no P fertilizer added | *** | 1.7 | *** | −1.3 |
| Rain-fed, 90 kg N $ha^{-1}$ and P fertilizer added versus no P fertilizer added | ns | 0.3 | ns | −0.2 |
| Irrigated, 90 kg N $ha^{-1}$ and K fertilizer added versus no K fertilizer added | ns | −0.5 | ns | −0.2 |
| Rain-fed, 90 kg N $ha^{-1}$ and K fertilizer added versus no K fertilizer added | ns | −0.1 | ** | −0.9 |

Standard error of the difference for the interaction of fertilizer treatment and water treatment

*, **, and *** are significant at the 0.1, 0.01, and 0.001 level, respectively

ns: not significant at the 0.1 level.

# DISCUSSION

The increased grain yield for plots that received a flush of irrigation and N fertilizer prior to planting in 2003, 2008, and 2011 is what was to be expected based upon others findings [3]. The adequate soil moisture at planting allowed for an improved stand establishment coupled with stimulated root and plant growth from the addition of N fertilizer. The interaction of irrigation and N fertilizer application timing was found not to be significant in 2004. This is likely because, in 2004, 131 mm of rain was received between late July and early October, the most for any of the four years analyzed (Figure 1). This amount of rainfall possibly negated the effect of pre plant irrigation on plant growth. The minimal increase in grain yields in 2003, 2008, and 2011 for rain-fed plots that received top dress N fertilizer could likely be due to the N fertilizer being applied to an actively growing crop with an established root system compared to that being applied prior to planting. There have been numerous conflicting results reported as to which is the most appropriate N application time to maximize grain yields in rain-fed winter wheat. The results that spring only N fertilizer applications have the potential to produce higher grain yields, reported in this study do agree with what has been reported by others such as Vaughan et al. [33].

As stated earlier, treatments that received the high N fertilizer rate prior to planting and P fertilizer exhibited higher grain yields in 2003 and 2008 (irrigated and/or rain-fed). Differences likely were not seen in 2004 because of ample precipitation and in 2011 because of severe drought (Figure 1). The only logical explanation for the significant increase in grain yield for the 2003 fertilizer treatments that received 45 kg N $ha^{-1}$ midseason compared to N applications prior to planting is that the period from planting to the average break of dormancy received the most amount of rainfall of any of the four years analyzed (Figure 1). The soil being near saturation during this period, confirmed by the elevated FWI values over this time (Figure 2), is ideal for N losses via denitrification [34].

The average soil test P values since the establishment of this trial for treatments that do not receive P fertilizer and do receive P fertilizer were 6 and 21 mg P $kg^{-1}$ for the irrigated site and 4 and 27 mg P $kg^{-1}$ for the rain-fed site, respectively (Table 1). Based on the soil test P values and the current Oklahoma State University recommendations [35], the sites that have received P fertilizer over the 45 plus years of this experiment have maintained soil test P values above sufficient levels. The treatments not receiving any P fertilizer have been at best 50 percent sufficient; thus it can be concluded when differences in P fertilizer response were detected, it was likely due more to the treatment being deficient in P and not so much the addition of P fertilizer.

A response to K fertilization on the soil type used in this study is not to be expected. The Hollister soil series has a clay mineralogy class denoted as semiotic [29]. Semiotic or montmorillonitic clays in semiarid regions are known for their shrink-swell properties and typically are high in K that is easily exchangeable or fixed within the clay lattices [36, 37]. The average soil test K values since the establishment of this trial for treatments that do not receive K fertilizer and do receive K fertilizer were 302 and 328 mg K $kg^{-1}$ for the irrigated site and 334 and 368 mg K $kg^{-1}$ for the rain-fed site, respectively. Treatments that have not received any K fertilizer in the 45 plus years of this experiment have soil test values well above the sufficient soil test K level that is recommended by Oklahoma State University [35]. The significant increase in grain yield in 2008 for the K fertilized treatments that received pre plant N compared to those that received top dress N could be related to the drier soil moisture conditions from planting until the break of dormancy. The improved or more distributed root growth from the addition of N fertilizer could have aided in the uptake of increased K and other nutrients during which was the driest period from planting to the average break of dormancy for the four years analyzed (Figure1).

When comparing the individual N, P, and K treatments that were either irrigated or rain-fed in 2003, 2008, and 2011, logically the irrigated treatments displayed higher grain yields. The response to P fertilizer was expressed for the 2003 and 2008 growing seasons and was more pronounced in the irrigated treatments compared to the rain-fed treatments. The only unexplainable trend when comparing the interaction of irrigation and fertilizer treatment was the significant reduction in grain yield in 2003 for irrigated treatments that received K fertilizer compared to plots not receiving K fertilizer. The only rational explanation is something besides treatments affected one of the plots and was not documented anywhere.

The 2011 growing season was characterized as the year that received the least rainfall of any of the four years analyzed (Figure 1). Trends of decreased yields with the addition of P and K fertilizers and the higher rate of N fertilizer applications were observed for this year. Though these yield decreases were minimal and sometimes not significant, they need to be accounted for. One possible explanation could be that there was a reduced stand and reduced early plant growth caused by salt injury from the fertilizer. Fertilizer injury from N, P, and K sources are well known and documented and low soil moisture content can greatly increase the potential of fertilizer damage [38–40].

The 2004 growing season was the only year evaluated that did not have any significant differences for the three-way or any of the two-way interactions. This is thought to be because there was more than sufficient rainfall provided throughout the growing season to negate any effect of pre plant irrigation (Figure 1). The 2004 crop year had the highest amount of total rainfall for any of the crop years evaluated in this study. The distribution of rainfall for the 2004 crop year was fairly evenly distributed with high amounts of precipitation prior to planting to ensure stand establishment and high amounts of rainfall coming around the break of winter dormancy and continuing on through times of rapid growth and nutrient uptake in the spring (Figure 1). This is also exhibited in the trends of the FWI over the growing season (Figure 2). The 2004 crop has the highest FWI at the time of planting, the two weeks prior to, and the two weeks after the top dress N fertilizer application.

The lack of differences between treatments that received P fertilizer compared to treatments that did not receive P fertilizer during a year of adequate soil moisture has been reported by other researchers [12–14] The logic behind this is that adequate water allows for increased root development, which leads to the plant's ability to acquire more of the native soil P. Producers in a rain-fed cropping system will not know at planting, when P fertilizer is applied, whether they will have sufficient moisture throughout the growing season to utilize more native soil P. Nonetheless, it would still be recommended that producers should soil test and apply P fertilizer to attain sufficient levels.

Because of the methodology used to calculate the WUE for each plot, most of the trends and differences in WUE were similar to the trends and differences observed in grain yield. The methodology used assumes that each plot utilizes the same amount of moisture added. To more accurately measure WUE for each treatment, the soil profile moisture concentration should be collected at both the beginning and the end of the growing season for each plot to know whether the crop utilized more or less soil moisture. One trend that was observed in WUE is that as N fertilizer's rate increased, as long as there is adequate moisture provided, the WUE increased as well and this has been documented by several other researchers [5, 6, 41]. Again, the logic is that with adequate moisture, the addition of N fertilizer stimulates root growth, which can easily promote the acquisition of more soil water and/or soil nutrients.

# CONCLUSIONS

Based upon the findings, whether irrigation or adequate soil moisture conditions are available prior to planting, applying sufficient N fertilizer pre plant is most beneficial to grain yield and WUE in the Central Rolling Red Plains. When the only source of moisture is through the naturally occurring rainfall and soil moisture conditions are average or below average, application timing of N fertilizer is not as important and can either be applied prior to planting or top dressed in the spring or potentially split between the two. The application of P fertilizer can be beneficial and producers should soil test and apply P fertilizer to achieve sufficient soil test levels to assist in optimizing grain yields. The application of K fertilizer is likely not beneficial in this region on these soil types and possibly even aided in the reduction of grain yields during dry years due to salt injury. In conclusion, having knowledge of the soil moisture content at certain times of the growing season, producers can better manage their fertilizer practices in the Central Rolling Red Plains.

## ACKNOWLEDGMENTS

The authors would like to thank the Oklahoma Soil Fertility Research and Education Advisory Board for their continued financial support of soil fertility research at Oklahoma State University. The authors would also like to thank the present and past field crew at the Oklahoma State University Southwest Research and Extension Center located near Altus, Oklahoma, for their maintenance of trials and documentation of agronomic activities.

## REFERENCES

1. U.S. Department of Agriculture, "Land Resource Regions and Major Land Areas of the United States, the Caribbean, and the Pacific Basin," USDA-NRCS, 2006,http://soils.usda.gov/survey/geography/mlra/index.html.
2. U.S. Department of Agriculture, Quick Stats. USDA-NASS, 2013,http://www.nass.usda.gov/Quick_Stats/index.php.
3. J. T. Musick and F. R. Lamm, "Pre plant irrigation in the central and southern high plains—a review," Transactions of the American Society of Agricultural Engineers, vol. 33, no. 6, pp. 1834–1842, 1990.
4. P. R. Gajri, S. S. Prihar, and V. K. Arora, "Effects of nitrogen and early irrigation on root development and water use by wheat on two soils," Field Crops Research, vol. 21, no. 2, pp. 103–114, 1989.
5. P. R. Gajri, S. S. Prihar, and V. K. Arora, "Interdependence of nitrogen and irrigation effects on growth and input-use efficiencies in wheat," Field Crops Research, vol. 31, no. 1-2, pp. 71–86, 1993.

6. G. Hussain and A. A. Al-Jaloud, "Effect of irrigation and nitrogen on water use efficiency of wheat in Saudi Arabia," Agricultural Water Management, vol. 27, no. 2, pp. 143–153, 1995.
7. K. P. Singh and V. Kumar, "Water use and water-use efficiency of wheat and barley in relation to seeding dates, levels of irrigation and nitrogen fertilization," Agricultural Water Management, vol. 3, no. 4, pp. 305–316, 1981.
8. R. K. Belford, B. Klepper, and R. W. Rickman, "Studies of intact root-shoot systems of field grown winter wheat—II. Root and shoot developmental patterns as related to nitrogen fertilizer," Agronomy Journal, vol. 79, pp. 310–319, 1987.
9. R. Singh, Y. Singh, S. S. Prihar, and P. Singh, "Effect of N fertilization on yield and water use efficiency of dryland winter wheat as affected by stored water and rainfall," Agronomy Journal, vol. 67, pp. 599–603, 1975.
10. D. Tennant, "Root growth of wheat—I. Early patterns of multiplication and extension of wheat roots including effects of levels of nitrogen, phosphorus, and potassium," Australian Journal of Agricultural Research, vol. 27, pp. 183–196, 1976.
11. S. R. Olsen, F. S. Watanabe, and R. E. Danielson, "Phosphorus absorption by corn roots as affected by moisture and phosphorus concentration," Proceedings of the Soil Science Society of America, vol. 25, pp. 289–294, 1961.
12. W. M. Strong and G. Barry, "The availability of soil and fertilizer phosphorus to wheat and rape at different water regimes," Australian Journal of Soil Research, vol. 18, pp. 353–362, 1980.
13. G. A. Reichman and D. L. Grunes, "Effect of water regime and fertilization on barley growth, water use, and N and P uptake," Agronomy Journal, vol. 58, pp. 513–517, 1966.
14. K. Simpson, "The significance of the effects of soil moisture and temperature on phosphorus uptake," in Soil Phosphorus, Tech. Bull. No. 13, pp. 19–29, Ministry of Agriculture, Fish. Food. H.M.S.O, London, UK, 1965.
15. R. Kuchenbuch, N. Claassen, and A. Jungk, "Potassium availability in relation to soil moisture," Plant and Soil, vol. 95, no. 2, pp. 233–243, 1986.

16. S. Seiffert, J. Kaselowsky, A. Jungk, and N. Claassen, "Observed and calculated potassium uptake by maize as affected by soil water content and bulk density," Agronomy Journal, vol. 87, no. 6, pp. 1070–1077, 1995.
17. Q. Zeng and P. H. Brown, "Soil potassium mobility and uptake by corn under differential soil moisture regimes," Plant and Soil, vol. 221, no. 2, pp. 121–134, 2000.
18. K. Mengel and L. C. Von Braunschweig, "The effect of soil moisture upon availability of potassium and its influence on the growth of young maize plants (Zea mays L.)," Soil Science, vol. 114, pp. 142–148, 1972.
19. J. M. Clarke, C. A. Campbell, H. W. Cutforth, R. M. DePauw, and G. E. Winkleman, "Nitrogen and phosphorus uptake, translocation, and utilization efficiency of wheat in relation to environment and cultivar yield and protein levels," Canadian Journal of Plant Science, vol. 70, pp. 965–977, 1990.
20. T. K. Das and N. T. Yaduraju, "Effect of weed competition on growth, nutrient uptake and yield of wheat as affected by irrigation and fertilizers," Journal of Agricultural Science, vol. 133, no. 1, pp. 45–51, 1999.
21. H. V. Eck, "Winter wheat response to nitrogen and irrigation," Agronomy Journal, vol. 80, pp. 902–908, 1988.
22. K. Girma, S. Holtz, B. Tubaña, J. Solie, and W. Raun, "Nitrogen accumulation in shoots as a function of growth stage of corn and winter wheat," Journal of Plant Nutrition, vol. 34, no. 2, pp. 165–182, 2011.
23. R. K. Boman, R. L. Westerman, W. R. Raun, and M. E. Jojola, "Time of nitrogen application: effects on winter wheat and residual soil nitrate," Soil Science Society of America Journal, vol. 59, no. 5, pp. 1364–1369, 1995.
24. M. M. Alcoz, F. M. Hons, and V. A. Haby, "Nitrogen fertilization timing effect on wheat production, nitrogen uptake efficiency, and residual soil nitrogen," Agronomy Journal, vol. 85, no. 6, pp. 1198–1203, 1993.
25. L. F. Welch, P. E. Johnson, J. W. Pendleton, and L. B. Miller, "Efficiency of fall- versus spring-applied nitrogen for winter wheat," Agronomy Journal, vol. 58, pp. 271–274, 1966.

26. L. A. Harper, R. R. Sharpe, G. W. Langdale, and J. E. Giddens, "Nitrogen cycling in a wheat crop: soil, plant, and aerial nitrogen transport," Agronomy Journal, vol. 79, pp. 965–973, 1987.
27. R. V. Olson and C. W. Shallow, "Fate of labeled nitrogen fertilizer applied to winter wheat,"Soil Science Society of America Journal, vol. 48, no. 3, pp. 583–586, 1984.
28. Soil Survey Staff, "Web Soil Survey: Soil Data Mart. USDA-NRCS," 2013,http://websoilsurvey.nrcs.usda.gov.
29. Soil Survey Staff, "Official Soil Series Description. USDA-NRCS," 2002,http://soils.usda.gov/technicalclassification/osd/index.html.
30. A. Mehlich, "Mehlich 3 soil test extractant: a modification of Mehlich 2 extractant,"Communications in Soil Science & Plant Analysis, vol. 15, no. 12, pp. 1409–1416, 1984.
31. Mesonet, Daily Data Retrieval. University of Oklahoma, 2013,http://www.mesonet.org/index.php/weather/category/past_data_files.
32. B. G. Illston, J. B. Basara, D. K. Fisher et al., "Mesoscale monitoring of soil moisture across a statewide network," Journal of Atmospheric and Oceanic Technology, vol. 25, no. 2, pp. 167–182, 2008.
33. B. Vaughan, D. G. Westfall, and K. A. Barbarick, "Nitrogen rate and timing effects on winter wheat grain yield, grain protein, and economics," Journal of Production Agriculture, vol. 3, pp. 324–328, 1990.
34. J. R. Burford and J. M. Bremner, "Relationships between the denitrification capacities of soils and total, water-soluble and readily decomposable soil organic matter," Soil Biology and Biochemistry, vol. 7, no. 6, pp. 389–394, 1975.
35. H. Zhang and W. R. Raun, Oklahoma Soil Fertility Handbook, Oklahoma State University Press, Stillwater, 6th edition, 2006.
36. J. L. Havlin, D. G. Westfall, and S. R. Olsen, "Mathematical models for potassium release kinetics in calcareous soils," Soil Science Society of America Journal, vol. 49, no. 2, pp. 371–376, 1985.
37. G. Sposito, The Chemistry of Soils, Oxford University Press, New York, NY, USA, 2nd edition, 2008.

38. O. G. Carter, "The effect of chemical fertilizer on seedling establishment," Australian Journal of Experimental Agriculture, vol. 7, pp. 174–180, 1967.

39. J. M. Scott, R. S. Jessop, R. J. Steer, and G. D. McLachlan, "Effect of nutrient seed coating on the emergence of wheat and oats," Fertilizer Research, vol. 14, no. 3, pp. 205–217, 1987.

40. C. K. Stevenson and T. E. Bates, "Effect of nitrogen to phosphorus atom ratio of ammonium phosphates on emergence of wheat (Triticum vulgare)," Agronomy Journal, vol. 60, pp. 493–495, 1968.

41. R. E. Ramig and H. F. Rhoades, "Interrelationships of soil moisture level at planting time and nitrogen fertilization on winter wheat production," Agronomy Journal, vol. 55, pp. 123–127, 1963.

Chapter 3

# Effect of Irrigation and Preplant Nitrogen Fertilizer Source on Maize in the Southern Great Plains

Jacob T. Bushong, Eric C. Miller, Jeremiah L. Mullock, D. Brian Arnall, and William R. Raun

Department of Plant and Soil Sciences, Oklahoma State University, 368 Agricultural Hall, Stillwater, OK 74078, USA

## ABSTRACT

With the demand for maize increasing, production has spread into more water limited, semiarid regions. Couple this with the increasing nitrogen (N) fertilizer costs and environmental concerns and the need for proper management practices has increased. A trial was established to evaluate the effects of different preplant N fertilizer sources on maize cultivated under deficit irrigation or rain-fed conditions on grain yield, N use efficiency (NUE), and water use efficiency (WUE). Two fertilizer sources, ammonium sulfate (AS) and urea ammonium nitrate (UAN), applied at two rates, 90 and 180 kg N ha$^{-1}$, were evaluated across four site-years. Deficit irrigation improved grain yield, WUE, and NUE compared to rain-fed conditions. The preplant application of a pure ammoniacal source of N fertilizer, such as AS, had a tendency to increase grain yields and NUE for rain-fed treatments. Under irrigated conditions, the use of UAN as a preplant N fertilizer source performed just as well or better at improving grain yield compared to AS, as long as the potential N loss mechanisms were minimized. Producers applying N preplant as a single application should adjust rates based on a reasonable yield goal and production practice.

# INTRODUCTION

Over the last two decades, the number of maize hectares planted and harvested in the Southern Great Plains of the United States has increased. While the number of irrigated hectares has remained fairly constant over this time span, the increase in rain-fed hectares has more than doubled [1]. This rise in area cultivated to maize is due to increased demand for maize for livestock feed exports and maize-based ethanol production [2]. With this increased production and an ever-growing concern for environmental implications, sustainable production practices that maximize the use of resources are being sought.

In some portions of the Southern Great Plains, groundwater is available for irrigation of maize production. However, in areas, such as the Ogallala Aquifer, the amount of water extracted from the aquifer has been much greater than the amount recharged leading to drastic declines in the water table which can exceed 50 percent of the saturated thickness [3]. One method utilized to better maximize maize grain yield and water use efficiency (WUE) has been deficit irrigation. Deficit irrigation is a management practice in which irrigation is applied below the evapotranspiration (ET) level at critical growth stages without significant reduction in grain yields [4]. The most critical growth stage at which moisture stress has been observed to be the most yield limiting in maize is the two weeks prior and the two weeks following silking [5]. Irrigation during the reproductive stages can still produce optimum grain yields and maximize WUE [6, 7].

The inefficient use of N fertilizer has been one of the major focal points for environmental contamination. A considerable factor affecting maize grain yield and N use efficiency (NUE) is the chemical make-up of the N fertilizer source. The source of the N fertilizer can impact the potential rate of loss and/or availability of the fertilizer [8]. According to Tsai et al. [9], utilizing ammoniacal-based N fertilizer sources may reduce potential losses via leaching and denitrification and may extend the availability of N in the soil for plant uptake throughout the growing season. Stevenson and Baldwin [10] compared the effects of ammonium nitrate, urea, and anhydrous ammonia applied at different times in maize. Regardless of application time, anhydrous ammonia yielded 240 to 260 kg $ha^{-1}$ more than both ammonium nitrate and urea. Power et al. [11] evaluated the effects of ammonium sulfate (AS), ammonium nitrate, calcium nitrate, and urea on maize grain yield and dry matter production. They reported that maize dry matter increased significantly with fertilization; however, grain yield differences among the different N sources were seldom significant. The ammoniacal sources typically displayed increased dry matter production with increasing N rates when compared to the calcium nitrate treatments while urea treatments were less than the other two ammoniacal sources. Olson et al. [12] compared anhydrous ammonia to urea ammonium nitrate (UAN) that was applied at planting or sidedress. They reported that anhydrous ammonia yielded more than the UAN treatment. They attributed the decreased yields in the UAN treatments to the nitrate component, which has the potential for being lost through leaching or denitrification, and the urea component, which has greater potential for N losses via ammonia volatilization. Freeman et al. [8] investigated the use of urea and anhydrous ammonia applied at different times with different soil incorporation procedures. They concluded that both grain yield and N uptake were improved when the N fertilizer source was urea, but only if the urea was applied and incorporated preplant or after harvest when residue incorporation is practiced.

The NUE and WUE of maize hybrids often coincide with one another [13] because of the greater response to N fertilizer with increases in added water [14, 15]. Because of this relationship, researchers have evaluated the effects of N fertilizer practices on WUE. For maize fields to be productive and resource-use efficient, numerous researchers have proposed a compromise of management practices that optimize grain yield and WUE. These practices include only applying N when water is adequate [15, 16], maintaining proper fertility based on tillage practices [16] and applying proper amounts of irrigation at critical growth stages [15, 17, 18].

The objectives of this study were to evaluate the interactive effects of two N fertilizer sources (UAN and AS), application rate, and deficit irrigation on maize early season vegetative growth, grain yield, NUE, and WUE. Our hypotheses for this trial are parallel to what previous researchers have documented in that the use of irrigation will increase not only the grain yield and NUE, but also the WUE. How efficient each fertilizer source will be for each system will be determined with the premise that the source that is more ammoniacally based will be more efficient regardless of system.

## MATERIALS AND METHODS

The experiment was conducted at two locations (Stillwater, OK, and Lake Carl Blackwell, OK) during the 2012 and 2013 growing seasons. Basic early spring preplant soil nutrient testing results (0–15 cm) and site soil mapping unit descriptions are provided in Table 1. If required, sites were fertilized prior to planting to 100 percent sufficient levels based on soil test P and K results and the fertilizer recommendations described in Zhang and Raun [19]. This practice was conducted to ensure that N was the only limiting nutrient.

**Table 1:** Preplant surface (0–15 cm) chemical characteristics and soil classification of sites utilized in this study

| **Location**[a] | **Year** | **Soil mapping unit** | **Major component soil taxonomic classification** | **pH**[b] | **$NH_4$-N**[c] | **$NO_3$-N**[c] | **$SO_4$-S**[d] | **P**[e] | **K**[e] | **Total N**[f] | **Organic C**[f] |
|---|---|---|---|---|---|---|---|---|---|---|---|
| | | | | | **mg kg$^{-1}$** | | | | | **g kg$^{-1}$** | |
| STW | 2012 | Easpur loam, 0 to 1 percent slopes, occasionally flooded | Easpur: fine-loamy, mixed, superactive, and thermic Fluventic Haplustolls | 6.2 | 11 | 4 | 13 | 30 | 119 | 0.8 | 9.4 |
| LCB | 2012 | Port-Oscar complex, 0 to 1 percent slopes, occasionally flooded | Port: fine-silty, mixed, superactive, and thermic Cumulic Haplustolls Oscar: fine-silty, mixed, superactive, and thermic Typic Nastrustalfs | 5.6 | 8 | 3 | 8 | 22 | 111 | 0.6 | 7.8 |
| STW | 2013 | Norge loam, 3 to 5 percent slopes | Norge: fine-silty, mixed, active, and thermic Udic Paleustolls | 5.0 | 16 | 11 | 15 | 87 | 117 | 1.2 | 10.5 |
| LCB | 2013 | Port-Oscar complex, 0 to 1 percent slopes, occasionally flooded | Port: fine-silty, mixed, superactive, and thermic Cumulic Haplustolls Oscar: fine-silty, mixed, superactive, and thermic Typic Nastrustalfs | 6.1 | 6 | 5 | 8 | 24 | 139 | 1.1 | 9.5 |

STW: Oklahoma State University Agriculture Experiment Station near Stillwater, OK; LCB: Oklahoma State University Agriculture Experiment Station near Lake Carl Blackwell, OK.

[b]1: 1 water.

[c]2M KCl extract [20].

[d]Calcium monophosphate extract [21].

[e]Mehlich III extract [22].

[f]Dry combustion [23].

A split-block experimental design with three replications per site-year was employed to evaluate the effects of irrigation and N fertilizer source in this experiment. Irrigated or rain-fed treatments served as the main plot, while five N fertilizer treatments based upon N source and N rate served as the subplot. Ammonium sulfate (AS, 21-0-0) and urea ammonium nitrate (UAN, 28-0-0) N fertilizer sources were evaluated in this experiment. Both fertilizer sources were applied at N rates of 90 and 180 kg N $ha^{-1}$. Fertilizer was broadcast applied and mechanically incorporated prior to planting. A complete list of the five N fertilizer treatments, which includes an unfertilized check, implemented to both irrigated and rain-fed plots, is provided in Table 2. To ensure that the added sulfur associated with the AS fertilizer would not have an effect on treatments, preplant soil samples were analyzed for sulfate-sulfur content (Table 1). The sulfate-sulfur soil test values were above sufficient levels described by the regional recommendations of Zhang and Raun [19].

**Table 2:** Nitrogen fertilizer treatment structure applied to both irrigated and rain-fed plots in this study

| Treatment number | Prplant N rate | Preplant N source[a] |
|---|---|---|
| | kg N $ha^{-1}$ | |
| 1 | 0 | — |
| 2 | 90 | UAN |
| 3 | 90 | AS |
| 4 | 180 | UAN |
| 5 | 180 | AS |

UAN: urea ammonium nitrate (28-0-0); AS: ammonium sulfate (21-0-0); applied prior to planting and mechanically incorporated.

For all site-years, plot sizes were 3.1 m wide by 6.2 m long. Four rows spaced at 76 cm apart were planted per plot and all measured observations were collected on the middle two rows. Field activities including planting dates, hybrids, seeding rates, N fertilizer application dates, irrigation totals, and harvest dates are provided in Table 3. Planting took place in the spring using maize hybrids that are known to express improved drought tolerance. Seeding rates were based on best agronomic practices for the region. The type of irrigation used was surface drip irrigation. Though this is not an economically viable option for irrigation in maize production, it was used strictly for research purposes. The use of drip irrigation allowed for the accurate measurement and placement of applied water. Two strips of drip tape were placed through each plot between the first and second rows and between the third and fourth rows. The amount of irrigation water (mm) distributed over each plot was determined by measuring the liters of water applied over the given area.

**Table 3:** Field activities for the four site-years utilized in this study

| **Field activity** | **2012** | | **2013** | |
|---|---|---|---|---|
| | **STW[a]** | **LCB[a]** | **STW** | **LCB** |
| Preplant N fertilization date | April 2 | April 5 | March 18 | March 18 |
| Planting date | April 9 | April 10 | March 20 | March 20 |
| Maize hybrid | Pioneer P1498HR | Pioneer P0876HR | Pioneer P1498HR | Dekalb 63–55 |
| Seeding rate (seeds $ha^{-1}$) | 49,000 | 49,000 | 54,000 | 54,000 |
| Start irrigation | May 16 | May 17 | June 13 | June 14 |
| Cease irrigation | July 11 | July 9 | July 9 | July 9 |
| Irrigation percent of PET[b] | 38 | 21 | 28 | 13 |
| Number of irrigations | 22 | 14 | 9 | 5 |
| Amount of irrigation (mm) | 173 | 89 | 55 | 27 |
| Amount of rainfall (mm) | 233 | 201 | 621 | 834 |
| Harvest date | August 6 | July 26 | September 9 | September 4 |

STW: Stillwater, OK; LCB: Lake Carl Blackwell, OK.[b]PET: potential evapotranspiration.

Potential differences in early vegetative growth/biomass accumulation were measured using the normalized difference vegetative index (NDVI) values collected with a Greenseeker (Trimble, Sunnyvale, CA, USA) ground based, optical sensor. Sensor readings were collected at the V6, V8, V10, and V12 growth stages [24] for all site-years.

Grain yield was determined by harvesting the center two rows of the four row plots with a Massey Ferguson 8XP self-propelled plot combine (Massey Ferguson, Duluth, GA, USA). Plot grain yields were adjusted for a standard moisture content of 155 $g\,kg^{-1}$. Grain subsamples were oven-dried and processed to pass a 140 mesh screen and were analyzed for total N content using a dry combustion analyzer. The NUE was then calculated by employing the difference method described by Varvel and Peterson [25] that utilizes the following equation:

$$\text{NUE} = \frac{(\text{Grain N uptake treated} - \text{Grain N uptake check})}{\text{N fertilizer added}} \quad (1)$$

Where grain N uptake for treated plots or the check plot was quantified by the percent N in the grain multiplied by the grain yield.

The WUE ($kg\,ha^{-1}\,mm^{-1}$) was measured for both site locations during the 2013 growing season. It was calculated as the ratio of dry grain yield ($kg\,ha^{-1}$) at 15.5 percent moisture to the seasonal water use expressed as ET. The ET was estimated using a mo

$$\text{ET} = \pm\Delta\text{SWC} + R + I, \quad (2)$$

Where ΔSWC is the change in soil profile (0 to 80 cm) volumetric soil water content from planting to harvest, R is the rainfall, I is the irrigation. It was assumed that water losses due to deep percolation or surface runoff were negligible. The ΔSWC was determined by collecting volumetric soil water samples from each plot with a 5 cm diameter probe long enough to encompass the 80 cm depth. The samples were collected using a hydraulic push probe (Giddings Machine Company, Windsor, CO, USA). Samples were collected the day prior to preplant fertilizer application and the day following grain harvest for each location. A moist weight was collected in the field and the samples were then oven-dried until no moisture was present in the sample. Daily rainfall was measured from the adjacent Oklahoma Mesonet [27] climate-monitoring station.

To understand the relationship of irrigation water applied to the daily potential ET (PET) for the trial area, daily PET values were determined. The PET values were derived from the American Society of Civil Engineers' Standardized Reference Evapotranspiration Equation described by Walter et al. [28]. Data collected as inputs for the equation to determine PET and rainfall were downloaded from the adjacent Oklahoma Mesonet [27] climate-monitoring site. The percent of irrigation water applied compared to PET losses for each site-year is described in Table 3.

Analysis of variance techniques was employed to detect significant differences for the main and interactive effects of treatments on early vegetative growth (NDVI), grain yield, NUE, and WUE. Single degree-of-freedom contrasts were used to partition statistical differences in treatment grouping means as well as detect any potential linear or quadratic trends based upon N fertilizer rate. All site-years were analyzed separately and thus the results are reported separately. For all analyses, an alpha level of 0.10 was used to determine statistical significance.

# RESULTS

## Stillwater, OK (2012)

### *Vegetative Growth*

No significant differences were observed in either the irrigated or rain-fed NDVI values for any of the growth stages evaluated (Figure 1). Regardless of treatment, the increase in NDVI appeared linear for the growth stages V6 through V10, and then plateaued between the V10 and V12 growth stages. One noticeable trend that was observed was that the 180 kg N $ha^{-1}$ UAN irrigated treatments had the lowest NDVI values for the V6, V8, and V10 growth stages, but the opposite was observed for that specific treatment under rain-fed conditions (Figure 1).

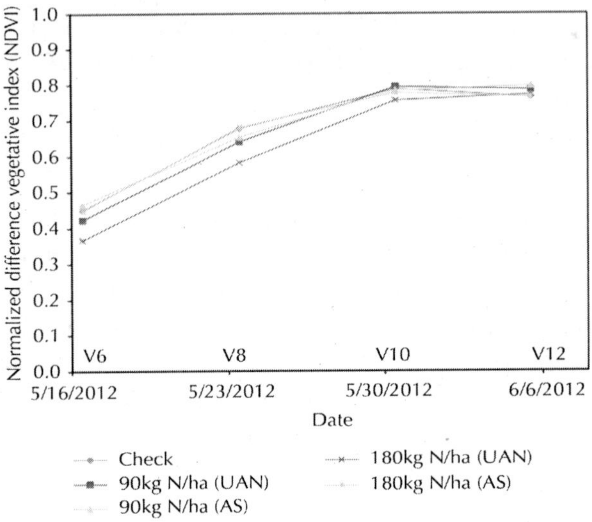


(a)

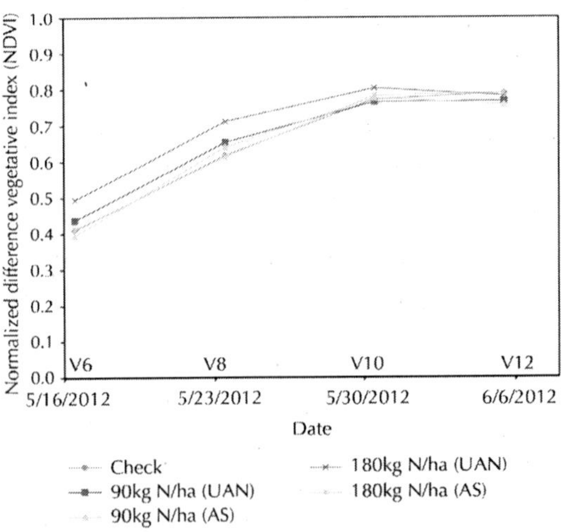


(b)

**Figure 1**: Normalized difference vegetative index (NDVI) values by maize growth stage for irrigated (a) and rain-fed (b) fertilizer treatments at Stillwater, OK (2012).

## Grain Yield

Irrigated and rain-fed grain yield values ranged from 6381 to 12265 kg ha$^{-1}$ and 2565 to 5980 kg ha$^{-1}$, respectively. Analysis of variance determined the effect of irrigation to be significant on grain yield (Table 4). On average, irrigated plots yielded about 4500 kg ha$^{-1}$ more than rain-fed plots (Table 5). The interactive effect of irrigation and fertilizer treatment was significant; however, the main effect of fertilizer treatment was not significant. Regardless of the fertilizer treatments being irrigated or rain-fed, AS treatments had numerically higher grain yields compared to the UAN treatments. This trend was also true for the rain-fed plots, except for the difference that was statistically significant (Table 6). Both the irrigated UAN and AS treatments displayed statistically significant linear increases in grain yield (Table 6). For rain-fed treatments, a quadratic trend was the only statistically significant N response trend for AS.

**Table 4:** P value results from the analysis of variance for the main and interactive effects of irrigation (Irr.) and preplant fertilizer treatment (Tmt.) on grain yield, N use efficiency (NUE), and water use efficiency (WUE)

| Source | Grain yield | NUE | WUE |
|---|---|---|---|
| STW[a] 2012 | | | |
| Irrigation | 0.0150 | 0.2258 | — |
| Treatment | 0.2241 | 0.6263 | — |
| Irr. × Tmt. | 0.0544 | 0.1089 | — |
| LCB[a] 2012 | | | |
| Irrigation | 0.0118 | 0.9156 | — |
| Treatment | 0.1355 | 0.0145 | — |
| Irr. × Tmt. | 0.3038 | 0.0394 | — |
| STW 2013 | | | |
| Irrigation | 0.0034 | 0.2243 | 0.0037 |
| Treatment | 0.0221 | 0.0381 | 0.0283 |
| Irr. × Tmt. | 0.1036 | 0.3306 | 0.1190 |
| LCB 2013 | | | |
| Irrigation | 0.0440 | 0.0415 | 0.0498 |
| Treatment | 0.0370 | 0.2215 | 0.0319 |
| Irr. × Tmt. | 0.5533 | 0.7275 | 0.4957 |

STW: Stillwater, OK; LCB: Lake Carl Blackwell, OK.

**Table 5:** Irrigated and rain-fed treatment means for grain yield, N use efficiency (NUE), and water use efficiency (WUE)

| Source | Grain yield | **NUE** | **WUE** |
|---|---|---|---|
| | **kg ha$^{-1}$** | % | **kg ha$^{-1}$ mm$^{-1}$** |
| STW[a] 2012 | | | |
| Irrigated | 8598 | 29.0 | — |
| Rain-fed | 4017 | 6.4 | — |
| P value | 0.0150 | 0.2258 | — |
| LCB[a] 2012 | | | |
| Irrigated | 6047 | 21.4 | — |
| Rain-fed | 4835 | 19.8 | — |
| P value | 0.0118 | 0.9156 | — |
| STW 2013 | | | |
| Irrigated | 9120 | 31.1 | 15.6 |
| Rain-fed | 2361 | 6.2 | 4.4 |
| P value | 0.0034 | 0.2243 | 0.0037 |
| LCB 2013 | | | |
| Irrigated | 8662 | 43.2 | 10.8 |
| Rain-fed | 4022 | 25.0 | 5.3 |
| P value | 0.0440 | 0.0415 | 0.0498 |

STW: Stillwater, OK; LCB: Lake Carl Blackwell, OK.

**Table 6:** Single degree-of-freedom contrast results for differences in treatment groupings for grain yield, N use efficiency (NUE), and water use efficiency (WUE) for Stillwater, OK (STW), in 2012 and 2013. Results listed under the "Main" column heading are the results of data pooled across irrigated and rain-fed treatments. Values in parenthesis are the difference in mean values for the group after the "versus" subtracted from the mean value of the group before the "versus"

| Contrast | Main | Irrigated | Rain-fed | Main | Irrigated | Rain-fed | Main | Irrigated | Rain-fed |
|---|---|---|---|---|---|---|---|---|---|
| | **Grain yield**[a] | | | **NUE**[a] | | | **WUE**[a] | | |
| STW 2012 | | | | | | | | | |
| UAN versus AS (difference) | ns[b] (−792) | ns (−241) | (−1343) | ns (0.4) | ns (4.8) | ns (−5.7) | — | — | — |
| UAN linear | ns | ** | ns | ns | ns | ns | — | — | — |
| UAN quadratic | ns | ns | ns | — | — | — | — | — | — |
| AS linear | ** | ** | ns | ns | * | ns | — | — | — |
| AS quadratic | ns | ns | * | — | — | — | — | — | — |
| STW 2013 | | | | | | | | | |
| UAN versus AS (difference) | (924) | (1463) | ns (385) | ns (4.8) | ns (8.4) | ns (1.3) | (1.4) | (2.1) | ns (0.7) |
| UAN linear | ** | *** | ns | ns | ns | ns | ** | *** | ns |
| UAN quadratic | ns | ns | ns | — | — | — | ns | ns | ns |
| AS linear | ns | ns | ns | ** | ** | ns | ns | ns | ns |
| AS quadratic | ** | ** | ns | — | — | — | ** | ** | ns |

Units: grain yield: $kg\,ha^{-1}$; NUE: percent; WUE: $kg\,ha^{-1}\,mm^{-1}$.

[b]ns: not significant at the 0.10 level.

[c]*,**,*** Significant at the 0.10, 0.05, and 0.01 level, respectively.

### *NUE*

Irrigated and rain-fed NUE values ranged from 5.6 to 60.7 percent and from nearly zero to 17.3 percent, respectively. Analysis of variance determined the effect of irrigation to be insignificant on NUE (Table 4). Though not statistically significant, irrigated plots improved NUE by more than 20 percent (Table 5). The analysis of variance did not detect significant differences for fertilizer treatment and the interaction of irrigation and fertilizer treatments. Single degree-of-freedom contrasts did not reveal any statistical differences in NUE between UAN and AS (Table 6). However, NUE values were numerically higher for the UAN irrigated treatments and NUE values were higher for the AS rain-fed treatments (Table 6). Because the check plots were used in the calculation of determining NUE, only linear trends could be observed. A negative linear trend was the only observed statistically significant trend for the AS treatments in the irrigated plots (Table 6).

## Lake Carl Blackwell, OK (2012)

### *Vegetative Growth*

No significant differences were observed in either the irrigated or rain-fed NDVI values for any of the growth stages evaluated (Figure 2). Regardless of treatment, the increase in NDVI was linear for growth stages V6 through V10 and then increased linearly between V10 and V12. One noticeable trend was that the unfertilized check treatments had the lowest NDVI values for the V8, V10, and V12 growth stages, but the opposite was observed for that specific treatment when rain-fed (Figure 2).

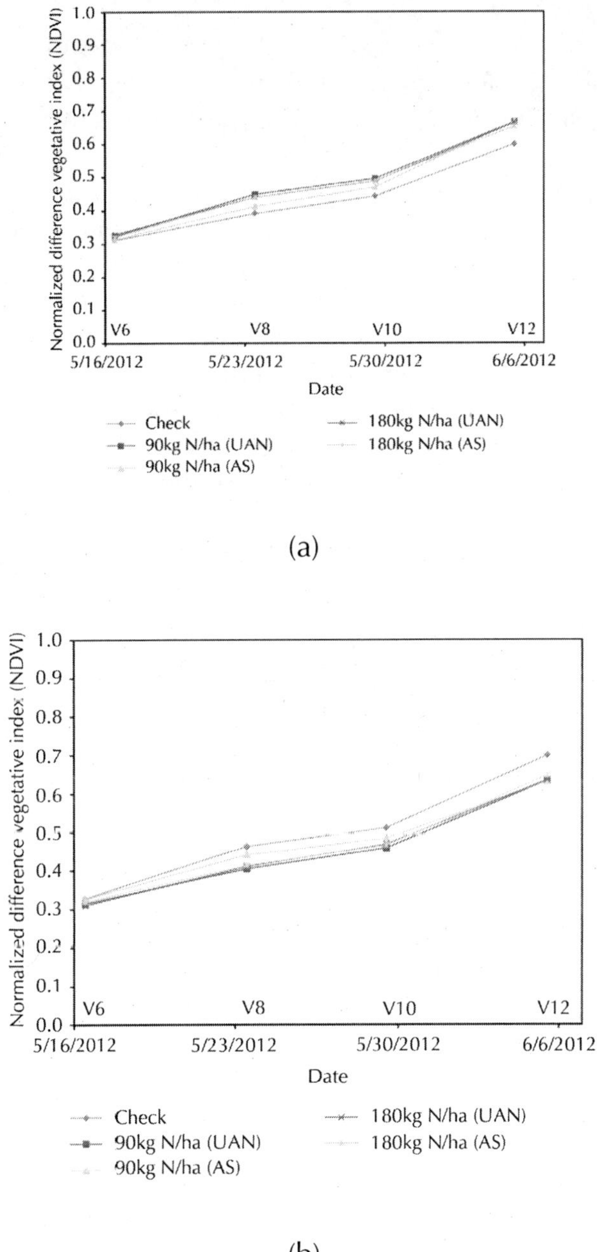


**Figure 2:** Normalized difference vegetative index (NDVI) values by maize growth stage for irrigated (a) and rain-fed (b) fertilizer treatments at Lake Carl Blackwell, OK (2012).

## Grain Yield

Irrigated and rain-fed grain yield values ranged from 4490 to 7351 kg $ha^{-1}$ and from 1322 to 6461 kg $ha^{-1}$, respectively. Analysis of variance determined a significant effect of irrigation on grain yield (Table 4). On average, irrigated plots yielded 1000 kg $ha^{-1}$ more than rain-fed plots (Table 5). No statistically significant differences were observed for fertilizer treatments and the interaction of irrigation and fertilizer treatments (Table 4). Regardless of the fertilizer treatments being irrigated or rain-fed, AS treatments had numerically higher grain yields compared to the UAN treatments (Table 7). No significant trends were observed for the response to UAN fertilizer (Table 7). A significant linear response was observed for AS in the irrigated plots and a quadratic response was observed for the rain-fed plots (Table 7).

**Table 7:** Single degree-of-freedom contrast results for differences in treatment groupings for grain yield, N use efficiency (NUE), and water use efficiency (WUE) for Lake Carl Blackwell, OK (LCB), in 2012 and 2013. Results listed under the "Main" column heading are the results of data pooled across irrigated and rain-fed treatments. Values in parenthesis are the difference in mean values for the group after the "versus" subtracted from the mean value of the group before the "versus"

| Contrast | Main | Irrigated | Rain-fed | Main | Irrigated | Rain-fed | Main | Irrigated | Rain-fed |
|---|---|---|---|---|---|---|---|---|---|
| | Grain yield[a] | | | **NUE[a]** | | | **WUE[a]** | | |
| LCB 2012 | | | | | | | | | |
| UAN versus AS (difference) | ns[b] (−303) | ns (−767) | ns (−569) | ns (−4.5) | Ns (1.0)** | (−10.0) | — | — | — |
| UAN linear | ns | ns | ns | ** | ** | ns | — | — | — |
| UAN quadratic | ns | ns | ns | — | — | — | — | — | — |
| AS linear | *[c] | * | ns | ** | ns | *** | — | — | — |
| AS quadratic | * | ns | ** | — | — | — | — | — | — |
| LCB 2013 | | | | | | | | | |
| UAN versus AS (difference) | ns (230) | ns (691) | ns (−232) | ns (2.2) | ns (6.0) | ns (−1.6) | ns (0.4) | ns (1.1) | ns (−0.4) |
| UAN linear | ** | ** | ns | * | ns | ns | ** | ** | ns |
| UAN quadratic | ns | * | ns | — | — | — | * | ** | ns |
| AS linear | ** | ** | ns | ns | ns | ns | ** | ** | ns |
| AS quadratic | ** | ns | ns | — | — | — | ns | ns | ns |

Units: grain yield: $kg\,ha^{-1}$; NUE: percent. WUE: $kg\,ha^{-1}\,mm^{-1}$.

[b]ns: not significant at the 0.10 level.

[c]*,**,*** Significant at the 0.10, 0.05, and 0.01 level, respectively.

### NUE

Irrigated and rain-fed NUE values ranged from 10.5 to 44.2 percent and from nearly zero to 78.3 percent, respectively. Analysis of variance determined the effect of irrigation to be insignificant on NUE (Table 4). When comparing irrigated versus rain-fed plots, no noticeable trend was observed in the differences between NUE values (Table 5). The analysis of variance did reveal significant differences for fertilizer treatment and the interaction of irrigation and fertilizer treatments. Regardless of the fertilizer treatments being irrigated or rain-fed, AS treatments displayed numerically higher NUE values (Table 7). This was especially true for the rain-fed plots in which the difference between UAN and AS was as much as 10 percent higher and was statistically significant (Table 7). Across irrigated and rain-fed treatments, significant, negatively linear responses were observed for both UAN and AS (Table 7). However, the linear response was only significant for UAN in the irrigated plots and AS in the rain-fed plots (Table 7).

## Stillwater, OK (2013)

### Vegetative Growth

Because irrigation did not commence until approximately the V12 or later growth stages, NDVI values were averaged across the irrigated and rain-fed treatments. No differences were observed for the V6, V10, and V12 growth stages; however, at the V8 growth stage, the NDVI value of the check treatment was significantly higher than the fertilized treatments (Figure 3). No distinct linear or quadratic trend was observed for the vegetative growth over time. The slopes of the lines between growth stages appeared to all be different, with the slope flattening out between the V10 and V12 growth stages (Figure 3).

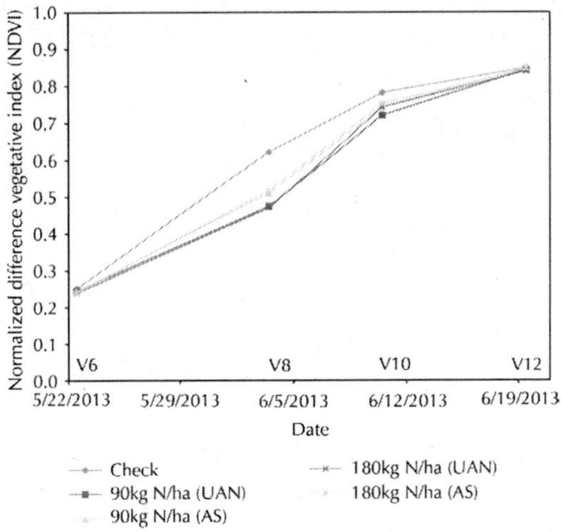


(a)

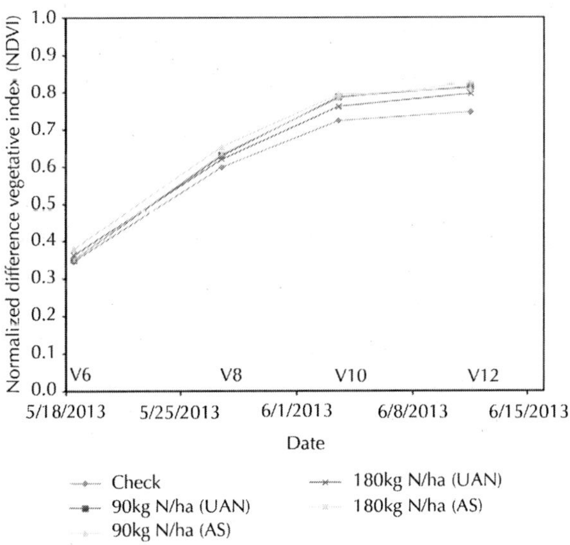


(b)

**Figure 3**: Normalized difference vegetative index (NDVI) values by maize growth stage, pooled across the main effect of irrigation, for Stillwater, OK (a) and Lake Carl Blackwell, OK (b) fertilizer treatments for the 2013 growing season.

## *Grain Yield*

Irrigated and rain-fed grain yield values ranged from 6020 to 11583 kg ha$^{-1}$ and 1345 to 3651 kg ha$^{-1}$, respectively. Analysis of variance determined a significant effect of irrigation on grain yield (Table 4). On average, irrigated plots yielded 6000 kg ha$^{-1}$ more than rain-fed plots (Table 5). No significant difference was observed for irrigation by fertilizer treatments interaction, but the effect of fertilizer treatments was observed to be significant (Table 4). Regardless of the fertilizer treatments applied to irrigated or rain-fed conditions, single degree-of-freedom contrasts revealed the response to UAN to be a linear response, whereas the response to AS was a quadratic response (Table 6). Overall, the UAN treatments significantly yielded more compared to AS fertilizer treatments. This was also true when fertilizer treatments were partitioned by irrigated and rain-fed treatments in which the differences in means were statistically and numerically higher, respectively (Table 6).

## *NUE*

Irrigated and rain-fed NUE values ranged from 6.5 to 83.7 percent and less than one to 25.4 percent, respectively. Analysis of variance determined the effect of irrigation to be insignificant on NUE, even though the average differences were greater than 20 percent (Table 4). The analysis of variance did reveal significant differences for fertilizer treatments, but not the interaction of irrigation and fertilizer treatments (Table 4). Single degree-of-freedom contrasts did not reveal any significant differences in NUE values between UAN and AS; however, the trend was that UAN gave numerically higher NUE values regardless of being irrigated or rain-fed (Table 6). No significant linear trend was observed for the UAN fertilizer treatments, but the AS treatments displayed a negative linear trend, especially for the irrigated treatments (Table 6).

## *WUE*

Irrigated and rain-fed WUE values ranged from 10.5 to 19.7 kg ha$^{-1}$ mm$^{-1}$ and 2.3 to 6.9 kg ha$^{-1}$ mm$^{-1}$, respectively. Analysis of variance determined the effect of irrigation to be significant on WUE (Table 4).

On average, irrigated plots yielded about 10 kg ha$^{-1}$ mm$^{-1}$ more than rain-fed plots (Table 5). The interactive effect of irrigation and fertilizer treatments was insignificant; however, the main effect of fertilizer treatment on WUE values was significant (Table 4). Single degree-of-freedom contrasts revealed UAN fertilizer treatments to be higher than AS treatments, which was significant regardless of irrigation treatment and the irrigated treatments (Table 6). Overall, the response to UAN tended to follow a linear trend, but the response to AS was a quadratic trend (Table 6).

## Lake Carl Blackwell, OK (2013)

### Vegetative Growth

As previously stated, since irrigation did not commence until approximately the V12 or later growth stages, NDVI values were averaged across the irrigated and rain-fed treatments. No significant differences in NDVI were observed between fertilizer treatments at any of the growth stages. Regardless of fertilizer treatment, the NDVI values tended to follow a quadratic pattern over time. One noticeable trend observed was that the check fertilizer plot had the lowest NDVI values for the V8, V10, and V12 growth stages (Figure 3).

### Grain Yield

Irrigated and rain-fed grain yield values ranged from 4675 to 12227 kg ha$^{-1}$ and 1327 to 6440 kg ha$^{-1}$, respectively. Analysis of variance determined a significant effect of irrigation on grain yield (Table 4). On average, irrigated plots yielded 4500 kg ha$^{-1}$ more than rain-fed plots (Table 5). No significant difference was observed for the irrigation by fertilizer treatments interaction, but the effect of fertilizer treatment was observed to be significant (Table 4). Even though it was not observed to be significant, AS treatments yielded higher than UAN treatments under rain-fed conditions with the opposite trend observed for the irrigated treatments (Table 7). A significant linear trend was observed for the UAN treatments regardless of being irrigated or rain-fed, but, for the AS treatments, significant linear and quadratic responses to fertilizer were observed (Table 7). Under irrigated conditions, the response of UAN treated plots was statically significant for both linear and quadratic trends, but only linear for the AS treatments (Table 7). No significant trends were observed for either N source under rain-fed conditions (Table 7).

### *NUE*

Irrigated and rain-fed NUE values ranged from 6.4 to 79.7 percent and from 2.7 to 70.7 percent, respectively. Analysis of variance determined the effect of irrigation to be significant for NUE values (Table 4). On average, irrigated plots yielded 20 percent more than rain-fed plots (Table 5). The analysis of variance did not reveal significant differences for fertilizer treatment, as well as the interaction of irrigation and fertilizer treatment (Table 4). Single degree-of-freedom contrasts revealed no significant trends or differences between fertilizer sources. In irrigated treatments, UAN had slightly numerically higher NUE values compared to AS; however, the opposite was observed for rain-fed conditions (Table 7).

### *WUE*

Irrigated and rain-fed WUE values ranged from 5.5 to 15.7 kg $ha^{-1}$ $mm^{-1}$ and 1.7 to 8.6 kg $ha^{-1}$ $mm^{-1}$, respectively. Analysis of variance determined the effect of irrigation to be significant on WUE (Table 4). On average, irrigated plots yielded about 5 kg $ha^{-1}$ $mm^{-1}$ more than rain-fed plots (Table 5). The interactive effect of irrigation and fertilizer treatments was insignificant; however, the main effect of fertilizer treatment on WUE values was significant (Table 4). No trends or differences were observed for either fertilizer source in the rain-fed areas (Table 7). No difference was observed in the WUE values between the UAN and AS treatments for the irrigated plots (Table 7). A significant quadratic response was observed for the UAN treatments; however, the highest ordered significant response for the AS treatments was a linear trend (Table 7).

## DISCUSSION

Deficit irrigation applied in the later vegetative and reproductive maize growth stages significantly increased grain yield and WUE. These results are what were to be expected.

Irrigation at times that have been deemed critical for optimum grain yield [5] has aided in optimizing yield [6,7]. Though it is only statistically significant for one of the four site-years, deficit irrigation also increased the NUE of the maize crop. Increases in NUE were likely due to greater N uptake and grain yield response to N fertilization. These results are similar to what has been observed by other researchers [14–17].

For three of the four site-years, rain-fed treatments had a greater yield response and increase in NUE for the AS treatments compared to the UAN treatments. This may be because of the more expansive root growth in the maize plant's attempt to acquire more soil moisture. The expansive root system would then have the ability to take up more of the immobile ammonium in the soil. Another desirable trait of ammoniacal N fertilizer sources in maize is that maize is able to take up ammonium during reproductive growth, whereas nitrate uptake is inhibited [9, 29]. Urea ammonium nitrate can be an effective N fertilizer source if the potential loss mechanisms (leaching, volatilization, and denitrification) are minimized [12]. The UAN treatments did outperform the AS treatments for the 2013 irrigated trials, but not the 2012 irrigated trials. This could be due to the fact that both 2013 sites had an above average rainfall for the region and, with adequate moisture early in the growing season, expansive root systems were not developed, which would have reduced ammonium acquisition from the soil. At Lake Carl Blackwell in 2013, the numerically lower yield response to UAN under irrigated conditions compared to the Stillwater site for that year and the lack of a numerical response to UAN for the rain-fed site could be due to potential N losses from the UAN. This site received the most rainfall of any site-year and we observed the topsoil to be saturated for a substantial amount of time prior to reproductive growth, thus leading to potential N losses via denitrification.

Little to no observable differences or trends in early season vegetative growth, as determined by collecting NDVI values, were present. However, with differences observed in grain yield and NUE between fertilizer treatments, there is the possibility that the inorganic N form (nitrate or ammonium) present in the soil later in the growing season affected grain yield and NUE.

To better optimize grain yield and NUE, the proper N fertilizer rate should be applied. The decrease in NUE values when the N fertilizer rate was increased from 90 to 180 kg N ha$^{-1}$ is typical for maize production and has been observed by others [8, 30]. For irrigated treatments, linear relationships with grain yield and N fertilizer rate were usually observed for both UAN and AS. However, a few of the rain-fed and irrigated site-years displayed statistically significant quadratic trends. These trends in which there is either a decrease or no increase in grain yield with added N above 90 kg N ha$^{-1}$ point towards excess N being applied and producers should adjust N application rates accordingly. With just two fertilizer rates plus a check treatment being employed, accurately determining an agronomic optimum preplant N fertilizer rate with the data from this trial would not be precise. However, producers should attempt to utilize some forms of a grain yield approach in making a preplant only N fertilizer rate recommendation or use regional N response trials from similar soil types under irrigation or rain-fed conditions.

Irrigated maize WUE values reportedly range from approximately 2 to 40 kg ha$^{-1}$ mm$^{-1}$ [31]. Irrigated WUE values observed in this experiment fell within this range. Variability in WUE values among treatments and growing seasons is to be expected. Zwart and Bastiaanssen [31] reported climate, water management, and soil fertility, all of which were evaluated in this trial and have the potential to give rise to the variability of WUE in maize. The main and interactive effects determined to be significant from the analysis of variance and single degree-of-freedom contrast results were similar for grain yield and WUE. This likely could be due to the manner in which WUE was calculated for this experiment, which involves the ratio of grain yield to the measured ET. One variable employed for deriving the ET was to measure the change in profile soil moisture prior to planting and immediately after harvest. Pre- and postharvest soil profile samples revealed no differences in the soil profile content between treatments (data not reported). The July and August months in the Southern Great Plains can be extremely hot and dry and likely much of the soil profile moisture was lost to evaporation and some transpiration during the grain dry-down period after irrigation had ceased. If no differences were observed in ET between fertilizer treatments within irrigated or rain-fed plots, then one can conclude that the differences in WUE would be dictated by the differences in grain yield.

# CONCLUSIONS

In conclusion, deficit irrigation during late vegetative and reproductive growth stages increased grain yield, NUE, and WUE. With three of the four rain-fed site-years reporting increases in grain yield and NUE, we would recommend that a pure ammoniacal N fertilizer source be applied if a preplant only N fertilizer application is to be utilized. If irrigation water is available, the N source is not as critical. However, the producer should be cognizant of the potential N loss mechanisms (leaching, volatilization, and denitrification) of N fertilizer sources like UAN. Lastly, if producers are going to utilize a preplant only fertilizer N application for maize cultivated on the Southern Great Plains, they should accordingly adjust N fertilizer rates based on a reasonable irrigated or rain-fed yield goal or regional N response trials.

# ACKNOWLEDGMENTS

The authors would like to thank the Oklahoma Soil Fertility Research and Education Advisory Board for their funding of this research project and their continued financial support of soil fertility research at Oklahoma State University.

# REFERENCES

1. U.S. Department of Agriculture, Quick Stats. USDA-NASS, 2014,http://www.nass.usda.gov/Quick_Stats/index.php.
2. S. Wallander, R. Claassen, and C. Nickerson, "The ethanol decade: an expansion of U.S. corn production, 2000–2009," Tech. Rep. EIB-79, U.S. Department of Agriculture, Economic Research Service, 2011.
3. M. Sophocleous, "Groundwater recharge and sustainability in the High Plains aquifer in Kansas, USA,"Hydrogeology Journal, vol. 13, no. 2, pp. 351–365, 2005.
4. FAO, "Defict irrigation practices," Water Reports 22, Food and Agriculture Organization of the United Nations, Rome, Italy, 2002.

5. B. R. Singh and D. P. Singh, "Agronomic and physiological responses of sorghum, maize, and pearl millet to irrigation," Field Crops Research, vol. 42, no. 2-3, pp. 57–67, 1995
6. R. K. Pandey, J. W. Maranville, and A. Admou, "Deficit irrigation and nitrogen effects on maize in a Sahelian environment I. Grain yield and yield components," Agricultural Water Management, vol. 46, no. 1, pp. 1–13, 2000.
7. S. Kang, W. Shi, and J. Zhang, "An improved water-use efficiency for maize grown under regulated deficit irrigation," Field Crops Research, vol. 67, no. 3, pp. 207–214, 2000
8. K. W. Freeman, K. Girma, R. K. Teal et al., "Long-term effects of nitrogen management practices on grain yield, nitrogen uptake, and efficiency in irrigated corn," Journal of Plant Nutrition, vol. 30, no. 12, pp. 2021–2036, 2007
9. C. Y. Tsai, I. Dweikat, D. M. Huber, and H. L. Warren, "Interrelationship of nitrogen nutrition with maize (Zea mays) grain yield, nitrogen use efficiency and grain quality," Journal of the Science of Food and Agriculture, vol. 58, no. 1, pp. 1–8, 1992.
10. C. K. Stevenson and C. S. Baldwin, "Effect of time and method of nitrogen application and source of nitrogen on the yield and nitrogen content of corn (Zea mays L.)," Agronomy Journal, vol. 61, pp. 381–384, 1969.
11. J. F. Power, J. Alessi, G. A. Reichman, and D. L. Grunes, "Effect of nitrogen source on corn and bromegrass production, soil pH, and inorganic soil nitrogen," Agronomy Journal, vol. 64, pp. 341–344, 1972
12. R. A. Olson, W. R. Raun, Y. S. Chun, and J. Skopp, "Nitrogen management and interseeding effects on irrigated corn and sorghum and on soil strength," Agronomy Journal, vol. 78, no. 5, pp. 856–862, 1986.
13. B. Eghball and J. W. Maranville, "Interactive effects of water and nitrogen stresses on nitrogen-utilization efficiency, leaf water status and yield of corn genotypes," Communications in Soil Science and Plant Analysis, vol. 22, pp. 1367–1382, 1991.
14. D. L. Martin, D. G. Watts, L. N. Mielke, K. D. Frank, and D. E. Eisenhauer, "Evaluation of nitrogen and irrigation management

for corn production using water high in nitrate," Soil Science Society of America Journal, vol. 46, pp. 1056–1062, 1982.

15. H. V. Eck, "Irrigated corn yield response to nitrogen and water," Agronomy Journal, vol. 76, pp. 421–428, 1984.
16. E. Di Paolo and M. Rinaldi, "Yield response of corn to irrigation and nitrogen fertilization in a Mediterranean environment," Field Crops Research, vol. 105, no. 3, pp. 202–210, 2008
17. M. M. Al-Kaisi and X. Yin, "Effects of nitrogen rate, irrigation rate, and plant population on corn yield and water use efficiency," Agronomy Journal, vol. 95, no. 6, pp. 1475–1482, 2003.
18. C. Mansouri-Far, S. A. M. M. Sanavy, and S. F. Saberali, "Maize yield response to deficit irrigation during low-sensitive growth stages and nitrogen rate under semi-arid climatic conditions," Agricultural Water Management, vol. 97, no. 1, pp. 12–22, 2010
19. H. Zhang and W. R. Raun, Oklahoma Soil Fertility Handbook, Oklahoma State University Press, Stillwater, Okla, USA, 6th edition, 2006.
20. R. L. Mulvaney, "Nitrogen-inorganic forms," in Methods of Soil Analysis. Part 3, D. L. Sparks, A. L. Page, P. A. Helmke, et al., Eds., Book Series 5, pp. 1123–1184, SSSA, Madison, Wis, USA, 1996.
21. M. A. Tabatabai, "Sulfur," in Methods of Soil Analysis. Part 3. Chemical Methods, L. D. Sparks, Ed., SSSA Book Series 5, pp. 921–960, SSSA, Madison, Wis, USA, 1996.
22. A. Mehlich, "Mehlich 3 soil test extractant: a modification of Mehlich 2 extractant," Communications in Soil Science & Plant Analysis, vol. 15, no. 12, pp. 1409–1416, 1984.
23. D. W. Nelson and L. E. Sommers, "Total carbon, organic carbon, and organic matter," in Methods of Soil Analysis: Part 3, D. L. Sparks, Soil Science Society of America, and American Society of Agronomy, Eds., SSSA Book Series 5, pp. 961–1010, SSSA, Madison, Wis, USA, 1996.
24. L. J. Abendroth, R. W. Elmore, M. J. Boyer, and S. K. Marlay, Corn Growth and Development, PMR 1009, Iowa State University Extension, Ames, Iowa, USA, 2011.
25. G. E. Varvel and T. A. Peterson, "Nitrogen fertilizer recovery by grain sorghum in monoculture and rotation systems," Agronomy Journal, vol. 83, pp. 617–622, 1991.

26. D. F. Heerman, "ET in irrigation managements," in Proceeding of the National Conference on Advances in Evapotranspiration, pp. 323–334, ASAE, 1985.

27. Mesonet, "Daily Data Retrieval," University of Oklahoma, 2014,http://www.mesonet.org/index.php/weather/category/past_data_files.

28. I. A. Walter, R. G. Allen, R. Elliot et al., "The ASCE standardized reference evapotranspiration equation," Tech. Rep., American Society of Civil Engineers, Environmental and Water Resources Institute, 2002.

29. W. L. Pan, E. J. Kamprath, R. H. Moll, and W. A. Jackson, "Prolificacy in corn: its effects on nitrate and ammonium uptake and utilization," Soil Science Society of America Journal, vol. 48, no. 5, pp. 1101–1106, 1984.

30. O. Walsh, W. Raun, A. Klatt, and J. Solie, "Effect of delayed nitrogen fertilization on maize (Zea mays L.) Grain yields and nitrogen use efficiency," Journal of Plant Nutrition, vol. 35, no. 4, pp. 538–555, 2012.

31. S. J. Zwart and W. G. M. Bastiaanssen, "Review of measured crop water productivity values for irrigated wheat, rice, cotton and maize," Agricultural Water Management, vol. 69, no. 2, pp. 115–133, 2004

Chapter 4

# Heavy Metals in Contaminated Soils: A Review of Sources, Chemistry, Risks and Best Available Strategies for Remediation

Raymond A. Wuana[1] and Felix E. Okieimen[2]

[1]Analytical Environmental Chemistry Research Group, Department of Chemistry, Benue State University, Makurdi 970001, Nigeria

[2]Research Laboratory, GeoEnvironmental & Climate Change Adaptation Research Centre, University of Benin, Benin City 300283, Nigeria

## ABSTRACT

Scattered literature is harnessed to critically review the possible sources, chemistry, potential biohazards and best available remedial strategies for a number of heavy metals (lead, chromium, arsenic, zinc, cadmium, copper, mercury and nickel) commonly found in contaminated soils. The principles, advantages and disadvantages of immobilization, soil washing and phytoremediation techniques which are frequently listed among the best demonstrated available technologies for cleaning up heavy metal contaminated sites are presented. Remediation of heavy metal contaminated soils is necessary to reduce the associated risks, make the land resource available for agricultural production, enhance food security and scale down land tenure problems arising from changes in the land use pattern.

# INTRODUCTION

Soils may become contaminated by the accumulation of heavy metals and metalloids through emissions from the rapidly expanding industrial areas, mine tailings, disposal of high metal wastes, leaded gasoline and paints, land application of fertilizers, animal manures, sewage sludge, pesticides, wastewater irrigation, coal combustion residues, spillage of petrochemicals, and atmospheric deposition [1, 2]. Heavy metals constitute an ill-defined group of inorganic chemical hazards, and those most commonly found at contaminated sites are lead (Pb), chromium (Cr), arsenic (As), zinc (Zn), cadmium (Cd), copper (Cu), mercury (Hg), and nickel (Ni) [3]. Soils are the major sink for heavy metals released into the environment by aforementioned anthropogenic activities and unlike organic contaminants which are oxidized to carbon (IV) oxide by microbial action, most metals do not undergo microbial or chemical degradation [4], and their total concentration in soils persists for a long time after their introduction [5]. Changes in their chemical forms (speciation) and bioavailability are, however, possible. The presence of toxic metals in soil can severely inhibit the biodegradation of organic contaminants [6]. Heavy metal contamination of soil may pose risks and hazards to humans and the ecosystem through: direct ingestion or contact with contaminated soil, the food chain (soil-plant-human or soil-plant-animal-human), drinking of contaminated ground water, reduction in food quality (safety and marketability) via phytotoxicity, reduction in land usability for agricultural production causing food insecurity, and land tenure problems [7–9].

The adequate protection and restoration of soil ecosystems contaminated by heavy metals require their characterization and remediation. Contemporary legislation respecting environmental protection and public health, at both national and international levels, are based on data that characterize chemical properties of environmental phenomena, especially those that reside in our food chain [10]. While soil characterization would provide an insight into heavy metal speciation and bioavailability, attempt at remediation of heavy metal contaminated soils would entail knowledge of the source of contamination, basic chemistry, and environmental and associated health effects (risks) of these heavy metals. Risk assessment is an effective scientific tool which enables decision makers to manage sites so contaminated in a cost-effective manner while preserving public and ecosystem health [11].

Immobilization, soil washing, and phytoremediation techniques are frequently listed among the best demonstrated available technologies (BDATs) for remediation of heavy metal-contaminated sites [3]. In spite of their cost-effectiveness and environment friendliness, field applications of these technologies have only been reported in developed countries. In most developing countries, these are yet to become commercially available technologies possibly due to the inadequate awareness of their inherent advantages and principles of operation. With greater awareness by the governments and the public of the implications of contaminated soils on human and animal health, there has been increasing interest amongst the scientific community in the development of technologies to remediate contaminated sites [12]. In developing countries with great population density and scarce funds available for environmental restoration, low-cost and ecologically sustainable remedial options are required to restore contaminated lands so as to reduce the associated risks, make the land resource available for agricultural production, enhance food security, and scale down land tenure problems.

In this paper, scattered literature is utilized to review the possible sources of contamination, basic chemistry, and the associated environmental and health risks of priority heavy metals (Pb, Cr, As, Zn, Cd, Cu, Hg, and Ni) which can provide insight into heavy metal speciation, bioavailability, and hence selection of appropriate remedial options. The principles, advantages, and disadvantages of immobilization, soil washing, and phytoremediation techniques as options for soil cleanup are also presented.

# SOURCES OF HEAVY METALS IN CONTAMINATED SOILS

Heavy metals occur naturally in the soil environment from the pedogenetic processes of weathering of parent materials at levels that are regarded as trace (<1000 mg $kg^{-1}$) and rarely toxic [10, 13]. Due to the disturbance and acceleration of nature's slowly occurring geochemical cycle of metals by man, most soils of rural and urban environments may accumulate one or more of the heavy metals above defined background values high enough to cause risks to human health, plants, animals, ecosystems, or other media [14]. The heavy metals essentially become contaminants in the soil environments because (i) their rates of generation via man-made cycles are more rapid relative to natural ones, (ii) they become transferred from mines to random environmental locations where higher potentials of direct exposure occur, (iii) the concentrations of the metals in discarded products are relatively high compared to those in the receiving environment, and (iv) the chemical form (species) in which a metal is found in the receiving environmental system may render it more bioavailable [14]. A simple mass balance of the heavy metals in the soil can be expressed as follows [15, 16]:

$$M_{total} = (M_p + M_a + M_f + M_{ow} + M_{ip}) - (M_{cr} + M_l), \quad (1)$$

where "$M$" is the heavy metal, "$p$" is the parent material, "$a$" is the atmospheric deposition, "$f$" is the fertilizer sources, "ag" are the agrochemical sources, "ow" are the organic waste sources, "ip" are other inorganic pollutants, "cr" is crop removal, and "$l$" is the losses by leaching, volatilization, and so forth. It is projected that the anthropogenic emission into the atmosphere, for several heavy metals, is one-to-three orders of magnitude higher than natural fluxes [17]. Heavy metals in the soil from anthropogenic sources tend to be more mobile, hence bioavailable than pedogenic, or lithogenic ones [18, 19]. Metal-bearing solids at contaminated sites can originate from a wide variety of anthropogenic sources in the form of metal mine tailings, disposal of high metal wastes in improperly protected landfills, leaded gasoline and lead-based paints, land application of fertilizer, animal manures, biosolids (sewage sludge), compost, pesticides, coal combustion residues, petrochemicals, and atmospheric deposition [1, 2, 20] are discussed hereunder.

## Fertilizers

Historically, agriculture was the first major human influence on the soil [21]. To grow and complete the lifecycle, plants must acquire not only macronutrients (N, P, K, S, Ca, and Mg), but also essential micronutrients. Some soils are deficient in the heavy metals (such as Co, Cu, Fe, Mn, Mo, Ni, and Zn) that are essential for healthy plant growth [22], and crops may be supplied with these as an addition to the soil or as a foliar spray. Cereal crops grown on Cu-deficient soils are occasionally treated with Cu as an addition to the soil, and Mn may similarly be supplied to cereal and root crops. Large quantities of fertilizers are regularly added to soils in intensive farming systems to provide adequate N, P, and K for crop growth. The compounds used to supply these elements contain trace amounts of heavy metals (e.g., Cd and Pb) as impurities, which, after continued fertilizer, application may significantly increase their content in the soil [23]. Metals, such as Cd and Pb, have no known physiological activity. Application of certain phosphatic fertilizers inadvertently adds Cd and other potentially toxic elements to the soil, including F, Hg, and Pb [24].

## Pesticides

Several common pesticides used fairly extensively in agriculture and horticulture in the past contained substantial concentrations of metals. For instance in the recent past, about 10% of the chemicals have approved for use as insecticides and fungicides in UK were based on compounds which contain Cu, Hg, Mn, Pb, or Zn. Examples of such pesticides are copper-containing fungicidal sprays such as Bordeaux mixture (copper sulphate) and copper oxychloride [23]. Lead arsenate was used in fruit orchards for many years to control some parasitic insects. Arsenic-containing compounds were also used extensively to control cattle ticks and to control pests in banana in New Zealand and Australia, timbers have been preserved with formulations of Cu, Cr, and As (CCA), and there are now many derelict sites where soil concentrations of these elements greatly exceed background concentrations. Such contamination has the potential to cause problems, particularly if sites are redeveloped for other agricultural or nonagricultural purposes. Compared with fertilizers, the use of such materials has been more localized, being restricted to particular sites or crops [8].

## Biosolids and Manures

The application of numerous biosolids (e.g., livestock manures, composts, and municipal sewage sludge) to land inadvertently leads to the accumulation of heavy metals such as As, Cd, Cr, Cu, Pb, Hg, Ni, Se, Mo, Zn, Tl, Sb, and so forth, in the soil [20]. Certain animal wastes such as poultry, cattle, and pig manures produced in agriculture are commonly applied to crops and pastures either as solids or slurries [25]. Although most manures are seen as valuable fertilizers, in the pig and poultry industry, the Cu and Zn added to diets as growth promoters and as contained in poultry health products may also have the potential to cause metal contamination of the soil [25, 26]. The manures produced from animals on such diets contain high concentrations of As, Cu, and Zn and, if repeatedly applied to restricted areas of land, can cause considerable buildup of these metals in the soil in the long run.

Biosolids (sewage sludge) are primarily organic solid products, produced by wastewater treatment processes that can be beneficially recycled [27]. Land application of biosolids materials is a common practice in many countries that allow the reuse of biosolids produced by urban populations [28]. The term sewage sludge is used in many references because of its wide recognition and its regulatory definition. However, the term biosolids is becoming more common as a replacement for sewage sludge because it is thought to reflect more accurately the beneficial characteristics inherent to sewage sludge [29]. It is estimated that in the United States, more than half of approximately 5.6 million dry tonnes of sewage sludge used or disposed of annually is land applied, and agricultural utilization of biosolids occurs in every region of the country. In the European community, over 30% of the sewage sludge is used as fertilizer in agriculture [29]. In Australia over 175 000 tonnes of dry biosolids are produced each year by the major metropolitan authorities, and currently most biosolids applied to agricultural land are used in arable cropping situations where they can be incorporated into the soil [8].

There is also considerable interest in the potential for composting biosolids with other organic materials such as sawdust, straw, or garden waste. If this trend continues, there will be implications for metal contamination of soils. The potential of biosolids for contaminating soils with heavy metals has caused great concern about their application in agricultural practices [30]. Heavy metals most commonly found in biosolids are Pb, Ni, Cd, Cr, Cu, and Zn, and the metal concentrations are governed by the nature and the intensity of the industrial activity, as well as the type of process employed during the biosolids treatment [31]. Under certain conditions, metals added to soils in applications of biosolids can be leached downwards through the soil profile and can have the potential to contaminate groundwater [32]. Recent studies on some New Zealand soils treated with biosolids have shown increased concentrations of Cd, Ni, and Zn in drainage leachates [33, 34].

## Wastewater

The application of municipal and industrial wastewater and related effluents to land dates back 400 years and now is a common practice in many parts of the world [35]. Worldwide, it is estimated that 20 million hectares of arable land are irrigated with waste water. In several Asian and African cities, studies suggest that agriculture based on wastewater irrigation accounts for 50 percent of the vegetable supply to urban areas [36]. Farmers generally are not bothered about environmental benefits or hazards and are primarily interested in maximizing their yields and profits. Although the metal concentrations in wastewater effluents are usually relatively low, long-term irrigation of land with such can eventually result in heavy metal accumulation in the soil.

## Metal Mining and Milling Processes and Industrial Wastes

Mining and milling of metal ores coupled with industries have bequeathed many countries, the legacy of wide distribution of metal contaminants in soil. During mining, tailings (heavier and larger particles settled at the bottom of the flotation cell during mining) are directly discharged into natural depressions, including onsite wetlands resulting in elevated concentrations [37]. Extensive Pb and zinc Zn ore mining and smelting have resulted in contamination of soil that poses risk to human and ecological health. Many reclamation methods used for these sites are lengthy and expensive and may not restore soil productivity. Soil heavy metal environmental risk to humans is related to bioavailability. Assimilation pathways include the ingestion of plant material grown in (food chain), or the direct ingestion (oral bioavailability) of, contaminated soil [38].

Other materials are generated by a variety of industries such as textile, tanning, petrochemicals from accidental oil spills or utilization of petroleum-based products, pesticides, and pharmaceutical facilities and are highly variable in composition. Although some are disposed of on land, few have benefits to agriculture or forestry. In addition, many are potentially hazardous because of their contents of heavy metals (Cr, Pb, and Zn) or toxic organic compounds and are seldom, if ever, applied to land. Others are very low in plant nutrients or have no soil conditioning properties [25].

## Air-Borne Sources

Airborne sources of metals include stack or duct emissions of air, gas, or vapor streams, and fugitive emissions such as dust from storage areas or waste piles. Metals from airborne sources are generally released as particulates contained in the gas stream. Some metals such as As, Cd, and Pb can also volatilize during high-temperature processing. These metals will convert to oxides and condense as fine particulates unless a reducing atmosphere is maintained [39]. Stack emissions can be distributed over a wide area by natural air currents until dry and/or wet precipitation mechanisms remove them from the gas stream. Fugitive emissions are often distributed over a much smaller area because emissions are made near the ground. In general, contaminant concentrations are lower in fugitive emissions compared to stack emissions. The type and concentration of metals emitted from both types of sources will depend on site-specific conditions. All solid particles in smoke from fires and in other emissions from factory chimneys are eventually deposited on land or sea; most forms of fossil fuels contain some heavy metals and this is, therefore, a form of contamination which has been continuing on a large scale since the industrial revolution began. For example, very high concentration of Cd, Pb, and Zn has been found in plants and soils adjacent to smelting works. Another major source of soil contamination is the aerial emission of Pb from the combustion of petrol containing tetraethyl lead; this contributes substantially to the content of Pb in soils in urban areas and in those adjacent to major roads. Zn and Cd may also be added to soils adjacent to roads, the sources being tyres, and lubricant oils [40].

# BASIC SOIL CHEMISTRY AND POTENTIAL RISKS OF HEAVY `METALS

The most common heavy metals found at contaminated sites, in order of abundance are Pb, Cr, As, Zn, Cd, Cu, and Hg [40]. Those metals are important since they are capable of decreasing crop production due to the risk of bioaccumulation and biomagnification in the food chain. There's also the risk of superficial and groundwater contamination. Knowledge of the basic chemistry, environmental, and associated health effects of these heavy metals is necessary in understanding their speciation, bioavailability, and remedial options. The fate and transport of a heavy metal in soil depends significantly on the chemical form and speciation of the metal. Once in the soil, heavy metals are adsorbed by initial fast reactions (minutes, hours), followed by slow adsorption reactions (days, years) and are, therefore, redistributed into different chemical forms with varying bioavailability, mobility, and toxicity [41, 42]. This distribution is believed to be controlled by reactions of heavy metals in soils such as (i) mineral precipitation and dissolution, (ii) ion exchange, adsorption, and desorption, (iii) aqueous complexation, (iv) biological immobilization and mobilization, and (v) plant uptake [43].

## Lead

Lead is a metal belonging to group IV and period 6 of the periodic table with atomic number 82, atomic mass 207.2, density 11.4 g $cm^{-3}$, melting point 327.4°C, and boiling point 1725°C. It is a naturally occurring, bluish-gray metal usually found as a mineral combined with other elements, such as sulphur (i.e., PbS, $PbSO_4$), or oxygen ($PbCO_3$), and ranges from 10 to 30 mg $kg^{-1}$ in the earth's crust [44]. Typical mean Pb concentration for surface soils worldwide averages 32 mg $kg^{-1}$ and ranges from 10 to 67 mg $kg^{-1}$ [10]. Lead ranks fifth behind Fe, Cu, Al, and Zn in industrial production of metals. About half of the Pb used in the U.S. goes for the manufacture of Pb storage batteries. Other uses include solders, bearings, cable covers, ammunition, plumbing, pigments, and caulking. Metals commonly alloyed with Pb are antimony (in storage batteries), calcium (Ca) and tin (Sn) (in maintenance-free storage batteries), silver (Ag) (for solder and anodes), strontium (Sr) and Sn (as anodes in electrowinning processes), tellurium (Te) (pipe and sheet in chemical installations and nuclear shielding), Sn (solders), and antimony (Sb), and Sn (sleeve bearings, printing, and high-detail castings) [45].

Ionic lead, Pb(II), lead oxides and hydroxides, and lead-metal oxyanion complexes are the general forms of Pb that are released into the soil, groundwater, and surface waters. The most stable forms of lead are Pb(II) and lead-hydroxy complexes. Lead(II) is the most common and reactive form of Pb, forming mononuclear and polynuclear oxides and hydroxides [3]. The predominant insoluble Pb compounds are lead phosphates, lead carbonates (form when the pH is above 6), and lead (hydr)oxides [46]. Lead sulfide (PbS) is the most stable solid form within the soil matrix and forms under reducing conditions, when increased concentrations of sulfide are present. Under anaerobic conditions a volatile organolead (tetramethyl lead) can be formed due to microbial alkylation [3].

Lead(II) compounds are predominantly ionic (e.g., $Pb^{2+}$ $SO_4^{2-}$), whereas Pb(IV) compounds tend to be covalent (e.g., tetraethyl lead, $Pb(C_2H_5)_4$). Some Pb (IV) compounds, such as $PbO_2$, are strong oxidants. Lead forms several basic salts, such as $Pb(OH)_2{\cdot}2PbCO_3$, which was once the most widely used white paint pigment and the source of considerable chronic lead poisoning to children who ate peeling white paint. Many compounds of Pb(II) and a few Pb(IV) compounds are useful. The two most common of these are lead dioxide and lead sulphate, which are participants in the reversible reaction that occurs during the charge and discharge of lead storage battery.

In addition to the inorganic compounds of lead, there are a number of organolead compounds such as tetraethyl lead. The toxicities and environmental effects of organolead compounds are particularly noteworthy because of the former widespread use and distribution of tetraethyllead as a gasoline additive. Although more than 1000 organolead compounds have been synthesized, those of commercial and toxicological importance are largely limited to the alkyl (methyl and ethyl) lead compounds and their salts (e.g., dimethyldiethyllead, trimethyllead chloride, and diethyllead dichloride).

Inhalation and ingestion are the two routes of exposure, and the effects from both are the same. Pb accumulates in the body organs (i.e., brain), which may lead to poisoning (plumbism) or even death. The gastrointestinal tract, kidneys, and central nervous system are also affected by the presence of lead. Children exposed to lead are at risk for impaired development, lower IQ, shortened attention span, hyperactivity, and mental deterioration, with children under the age of six being at a more substantial risk. Adults usually experience decreased reaction time, loss of memory, nausea, insomnia, anorexia, and weakness of the joints when exposed to lead [47]. Lead is not an essential element. It is well known to be toxic and its effects have been more extensively reviewed than the effects of other trace metals. Lead can cause serious injury to the brain, nervous system, red blood cells, and kidneys [48]. Exposure to lead can result in a wide range of biological effects depending on the level and duration of exposure. Various effects occur over a broad range of doses, with the developing young and infants being more sensitive than adults. Lead poisoning, which is so severe as to cause evident illness, is now very rare. Lead performs no known essential function in the human body, it can merely do harm after uptake from food, air, or water. Lead is a particularly dangerous chemical, as it can accumulate in individual organisms, but also in entire food chains.

The most serious source of exposure to soil lead is through direct ingestion (eating) of contaminated soil or dust. In general, plants do not absorb or accumulate lead. However, in soils testing high in lead, it is possible for some lead to be taken up. Studies have shown that lead does not readily accumulate in the fruiting parts of vegetable and fruit crops (e.g., corn, beans, squash, tomatoes, strawberries, and apples). Higher concentrations are more likely to be found in leafy vegetables (e.g., lettuce) and on the surface of root crops (e.g., carrots). Since plants do not take up large quantities of soil lead, the lead levels in soil considered safe for plants will be much higher than soil lead levels where eating of soil is a concern (pica). Generally, it has been considered safe to use garden produce grown in soils with total lead levels less than 300 ppm. The risk of lead poisoning through the food chain increases as the soil lead level rises above this concentration. Even at soil levels above 300 ppm, most of the risk is from lead contaminated soil or dust deposits on the plants rather than from uptake of lead by the plant [49].

## Chromium

Chromium is a first-row *d*-block transition metal of group VIB in the periodic table with the following properties: atomic number 24, atomic mass 52, density 7.19 g $cm^{-3}$, melting point 1875°C, and boiling point 2665°C. It is one of the less common elements and does not occur naturally in elemental form, but only in compounds. Chromium is mined as a primary ore product in the form of the mineral chromite, $FeCr_2O_4$. Major sources of Cr-contamination include releases from electroplating processes and the disposal of Cr containing wastes [39]. Chromium(VI) is the form of Cr commonly found at contaminated sites. Chromium can also occur in the +III oxidation state, depending on pH and redox conditions. Chromium(VI) is the dominant form of Cr in shallow aquifers where aerobic conditions exist. Chromium(VI) can be reduced to Cr(III) by soil organic matter, $S^{2-}$ and $Fe^{2+}$ ions under anaerobic conditions often encountered in deeper groundwater. Major Cr(VI) species include chromate ($CrO_4^{2-}$) and dichromate ($Cr_2O_7^{2-}$) which precipitate readily in the presence of metal cations (especially $Ba^{2+}$, $Pb^{2+}$, and $Ag^{+}$). Chromate and dichromate also adsorb on soil surfaces, especially iron and aluminum oxides. Chromium(III) is the dominant form of Cr at low pH (<4). $Cr^{3+}$ forms solution complexes with $NH_3$, $OH^-$, $Cl^-$, $F^-$, $CN^-$, $SO_4^{2-}$, and soluble organic ligands. Chromium(VI) is the more toxic form of chromium and is also more mobile. Chromium(III) mobility is decreased by adsorption to clays and oxide minerals below pH 5 and low solubility above pH 5 due to the formation of $Cr(OH)_3(s)$ [50]. Chromium mobility depends on sorption characteristics of the soil, including clay content, iron oxide content, and the amount of organic matter present. Chromium can be transported by surface runoff to surface waters in its soluble or precipitated form. Soluble and un-adsorbed chromium complexes can leach from soil into groundwater. The leachability of Cr(VI) increases as soil pH increases. Most of Cr released into natural waters is particle associated, however, and is ultimately deposited into the sediment [39]. Chromium is associated with allergic dermatitis in humans [21].

## Arsenic

Arsenic is a metalloid in group VA and period 4 of the periodic table that occurs in a wide variety of minerals, mainly as $As_2O_3$, and can be recovered from processing of ores containing mostly Cu, Pb, Zn, Ag and Au. It is also present in ashes from coal combustion. Arsenic has the following properties: atomic number 33, atomic mass 75, density 5.72 g $cm^{-3}$, melting point 817°C, and boiling point 613°C, and exhibits fairly complex chemistry and can be present in several oxidation states (−III, 0, III, V) [39]. In aerobic environments, As (V) is dominant, usually in the form of arsenate ($AsO_4^{3-}$) in various protonation states: $H_3AsO_4$, $H_2AsO_4^-$, $HAsO_4^{2-}$, and $AsO_4^{3-}$. Arsenate and other anionic forms of arsenic behave as chelates and can precipitate when metal cations are present [51]. Metal arsenate complexes are stable only under certain conditions. Arsenic (V) can also coprecipitate with or adsorb onto iron oxyhydroxides under acidic and moderately reducing conditions. Coprecipitates are immobile under these conditions, but arsenic mobility increases as pH increases [39]. Under reducing conditions As(III) dominates, existing as arsenite ($AsO_3^{3-}$), and its protonated forms $H_3AsO_3$, $H_2AsO_3^-$, and $HAsO_3^{2-}$. Arsenite can adsorb or coprecipitate with metal sulfides and has a high affinity for other sulfur compounds. Elemental arsenic and arsine, $AsH_3$, may be present under extreme reducing conditions. Biotransformation (via methylation) of arsenic creates methylated derivatives of arsine, such as dimethyl arsine $HAs(CH_3)_2$ and trimethylarsine $As(CH_3)_3$ which are highly volatile. Since arsenic is often present in anionic form, it does not form complexes with simple anions such as $Cl^-$ and $SO_4^{2-}$. Arsenic speciation also includes organometallic forms such as methylarsinic acid $(CH_3)AsO_2H_2$ and dimethylarsinic acid $(CH_3)_2AsO_2H$. Many As compounds adsorb strongly to soils and are therefore transported only over short distances in groundwater and surface water. Arsenic is associated with skin damage, increased risk of cancer, and problems with circulatory system [21].

## Zinc

Zinc is a transition metal with the following characteristics: period 4, group IIB, atomic number 30, atomic mass 65.4, density 7.14 g $cm^{-3}$, melting point 419.5°C, and boiling point 906°C. Zinc occurs naturally in soil (about 70 mg $kg^{-1}$ in crustal rocks) [52], but Zn concentrations are rising unnaturally, due to anthropogenic additions. Most Zn is added during industrial activities, such as mining, coal, and waste combustion and steel processing. Many foodstuffs contain certain concentrations of Zn. Drinking water also contains certain amounts of Zn, which may be higher when it is stored in metal tanks. Industrial sources or toxic waste sites may cause the concentrations of Zn in drinking water to reach levels that can cause health problems. Zinc is a trace element that is essential for human health. Zinc shortages can cause birth defects. The world's Zn production is still on the rise which means that more and more Zn ends up in the environment. Water is polluted with Zn, due to the presence of large quantities present in the wastewater of industrial plants. A consequence is that Zn-polluted sludge is continually being deposited by rivers on their banks. Zinc may also increase the acidity of waters. Some fish can accumulate Zn in their bodies, when they live in Zn-contaminated waterways. When Zn enters the bodies of these fish, it is able to biomagnify up the food chain. Water-soluble zinc that is located in soils can contaminate groundwater. Plants often have a Zn uptake that their systems cannot handle, due to the accumulation of Zn in soils. Finally, Zn can interrupt the activity in soils, as it negatively influences the activity of microorganisms and earthworms, thus retarding the breakdown of organic matter [53].

## Cadmium

Cadmium is located at the end of the second row of transition elements with atomic number 48, atomic weight 112.4, density 8.65 g $cm^{-3}$, melting point 320.9°C, and boiling point 765°C. Together with Hg and Pb, Cd is one of the big three heavy metal poisons and is not known for any essential biological function. In its compounds, Cd occurs as the divalent Cd(II) ion. Cadmium is directly below Zn in the periodic table and has a chemical similarity to that of Zn, an essential micronutrient for plants and animals. This may account in part for Cd's toxicity; because Zn being an essential trace element, its substitution by Cd may cause the malfunctioning of metabolic processes [54].

The most significant use of Cd is in Ni/Cd batteries, as rechargeable or secondary power sources exhibiting high output, long life, low maintenance, and high tolerance to physical and electrical stress. Cadmium coatings provide good corrosion resistance coating to vessels and other vehicles, particularly in high-stress environments such as marine and aerospace. Other uses of cadmium are as pigments, stabilizers for polyvinyl chloride (PVC), in alloys and electronic compounds. Cadmium is also present as an impurity in several products, including phosphate fertilizers, detergents and refined petroleum products. In addition, acid rain and the resulting acidification of soils and surface waters have increased the geochemical mobility of Cd, and as a result its surface-water concentrations tend to increase as lake water pH decreases [54]. Cadmium is produced as an inevitable byproduct of Zn and occasionally lead refining. The application of agricultural inputs such as fertilizers, pesticides, and biosolids (sewage sludge), the disposal of industrial wastes or the deposition of atmospheric contaminants increases the total concentration of Cd in soils, and the bioavailability of this Cd determines whether plant Cd uptake occurs to a significant degree [28]. Cadmium is very biopersistent but has few toxicological properties and, once absorbed by an organism, remains resident for many years.

Since the 1970s, there has been sustained interest in possible exposure of humans to Cd through their food chain, for example, through the consumption of certain species of shellfish or vegetables. Concern regarding this latter route (agricultural crops) led to research on the possible consequences of applying sewage sludge (Cd-rich biosolids) to soils used for crops meant for human consumption, or of using cadmium-enriched phosphate fertilizer [54]. This research has led to the stipulation of highest permissible concentrations for a number of food crops [8].

Cadmium in the body is known to affect several enzymes. It is believed that the renal damage that results in proteinuria is the result of Cd adversely affecting enzymes responsible for reabsorption of proteins in kidney tubules. Cadmium also reduces the activity of delta-aminolevulinic acid synthetase, arylsulfatase, alcohol dehydrogenase, and lipoamide dehydrogenase, whereas it enhances the activity of delta-aminolevulinic acid dehydratase, pyruvate dehydrogenase, and pyruvate decarboxylase [45]. The most spectacular and publicized occurrence of cadmium poisoning resulted from dietary intake of cadmium by people in the Jintsu River Valley, near Fuchu, Japan. The victims were afflicted by itai itai disease, which means ouch, ouch in

Japanese. The symptoms are the result of painful osteomalacia (bone disease) combined with kidney malfunction. Cadmium poisoning in the Jintsu River Valley was attributed to irrigated rice contaminated from an upstream mine producing Pb, Zn, and Cd. The major threat to human health is chronic accumulation in the kidneys leading to kidney dysfunction. Food intake and tobacco smoking are the main routes by which Cd enters the body [45].

## Copper

Copper is a transition metal which belongs to period 4 and group IB of the periodic table with atomic number 29, atomic weight 63.5, density 8.96 g $cm^{-3}$, melting point 1083°C and boiling point 2595°C. The metal's average density and concentrations in crustal rocks are 8.1 $\times 10^3$ kg $m^{-3}$ and 55 mg $kg^{-1}$, respectively [52].

Copper is the third most used metal in the world [55]. Copper is an essential micronutrient required in the growth of both plants and animals. In humans, it helps in the production of blood haemoglobin. In plants, Cu is especially important in seed production, disease resistance, and regulation of water. Copper is indeed essential, but in high doses it can cause anaemia, liver and kidney damage, and stomach and intestinal irritation. Copper normally occurs in drinking water from Cu pipes, as well as from additives designed to control algal growth. While Cu's interaction with the environment is complex, research shows that most Cu introduced into the environment is, or rapidly becomes, stable and results in a form which does not pose a risk to the environment. In fact, unlike some man-made materials, Cu is not magnified in the body or bioaccumulated in the food chain. In the soil, Cu strongly complexes to the organic implying that only a small fraction of copper will be found in solution as ionic copper, Cu(II). The solubility of Cu is drastically increased at pH 5.5 [56], which is rather close to the ideal farmland pH of 6.0–6.5 [57].

Copper and Zn are two important essential elements for plants, microorganisms, animals, and humans. The connection between soil and water contamination and metal uptake by plants is determined by many chemical and physical soil factors as well as the physiological properties of the crops. Soils contaminated with trace metals may pose both direct and indirect threats: direct, through negative effects of metals on crop growth and yield, and indirect, by entering the human food chain with a potentially negative impact on human health. Even a reduction of crop yield by a few percent could lead to a significant long-term loss in production and income. Some food importers are now specifying acceptable maximum contents of metals in food, which might limit the possibility for the farmers to export their contaminated crops [36].

## Mercury

Mercury belongs to same group of the periodic table with Zn and Cd. It is the only liquid metal at stp. It has atomic number 80, atomic weight 200.6, density 13.6 g cm$^{-3}$, melting point −13.6°C, and boiling point 357°C and is usually recovered as a byproduct of ore processing [39]. Release of Hg from coal combustion is a major source of Hg contamination. Releases from manometers at pressure-measuring stations along gas/oil pipelines also contribute to Hg contamination. After release to the environment, Hg usually exists in mercuric ($Hg^{2+}$), mercurous ($Hg_2^{2+}$), elemental ($Hg^o$), or alkylated form (methyl/ethyl mercury). The redox potential and pH of the system determine the stable forms of Hg that will be present. Mercurous and mercuric mercury are more stable under oxidizing conditions. When mildly reducing conditions exist, organic or inorganic Hg may be reduced to elemental Hg, which may then be converted to alkylated forms by biotic or abiotic processes. Mercury is most toxic in its alkylated forms which are soluble in water and volatile in air [39]. Mercury(II) forms strong complexes with a variety of both inorganic and organic ligands, making it very soluble in oxidized aquatic systems [51]. Sorption to soils, sediments, and humic materials is an important mechanism for the removal of Hg from solution. Sorption is pH dependent and increases as pH increases. Mercury may also be removed from solution by coprecipitation with sulphides. Under anaerobic conditions, both organic and inorganic forms of Hg may be converted to alkylated forms by microbial activity, such as by sulfur-reducing bacteria. Elemental mercury may also be formed under anaerobic conditions by demethylation of methyl mercury, or by reduction of Hg(II). Acidic conditions ($pH < 4$) also favor the formation of methyl mercury, whereas higher pH values favor precipitation of HgS(s) [39]. Mercury is associated with kidney damage [21].

## Nickel

Nickel is a transition element with atomic number 28 and atomic weight 58.69. In low pH regions, the metal exists in the form of the nickelous ion, Ni(II). In neutral to slightly alkaline solutions, it precipitates as nickelous hydroxide, $Ni(OH)_2$, which is a stable compound. This precipitate readily dissolves in acid solutions forming Ni(III) and in very alkaline conditions; it forms nickelite ion, $HNiO_2$, that is soluble in water. In very oxidizing and alkaline conditions, nickel exists in form of the stable nickelo-nickelic oxide, $Ni_3O_4$, that is soluble in acid solutions. Other nickel oxides such as nickelic oxide, $Ni_2O_3$, and nickel peroxide, $NiO_2$, are unstable in alkaline solutions and decompose by giving off oxygen. In acidic regions, however, these solids dissolve producing $Ni^{2+}$ [58].

Nickel is an element that occurs in the environment only at very low levels and is essential in small doses, but it can be dangerous when the maximum tolerable amounts are exceeded. This can cause various kinds of cancer on different sites within the bodies of animals, mainly of those that live near refineries. The most common application of Ni is an ingredient of steel and other metal products. The major sources of nickel contamination in the soil are metal plating industries, combustion of fossil fuels, and nickel mining and electroplating [59]. It is released into the air by power plants and trash incinerators and settles to the ground after undergoing precipitation reactions. It usually takes a long time for nickel to be removed from air. Nickel can also end up in surface water when it is a part of wastewater streams. The larger part of all Ni compounds that are released to the environment will adsorb to sediment or soil particles and become immobile as a result. In acidic soils, however, Ni becomes more mobile and often leaches down to the adjacent groundwater. Microorganisms can also suffer from growth decline due to the presence of Ni, but they usually develop resistance to Ni after a while. Nickel is not known to accumulate in plants or animals and as a result Ni has not been found to biomagnify up the food chain. For animals Ni is an essential foodstuff in small amounts. The primary source of mercury is the sulphide ore cinnabar.

# SOIL CONCENTRATION RANGES AND REGULATORY GUIDELINES FOR SOME HEAVY METALS

The specific type of metal contamination found in a contaminated soil is directly related to the operation that occurred at the site. The range of contaminant concentrations and the physical and chemical forms of contaminants will also depend on activities and disposal patterns for contaminated wastes on the site. Other factors that may influence the form, concentration, and distribution of metal contaminants include soil and ground-water chemistry and local transport mechanisms [3].

Soils may contain metals in the solid, gaseous, or liquid phases, and this may complicate analysis and interpretation of reported results. For example, the most common method for determining the concentration of metals contaminants in soil is via total elemental analysis (USEPA Method 3050). The level of metal contamination determined by this method is expressed as mg metal $kg^{-1}$ soil. This analysis does not specify requirements for the moisture content of the soil and may therefore include soil water. This measurement may also be reported on a dry soil basis. The level of contamination may also be reported as leachable metals as determined by leach tests, such as the toxicity characteristic leaching procedure (TCLP) (USEPA Method 1311) or the synthetic precipitation-leaching procedure, or SPLP test (USEPA Method 1312). These procedures measure the concentration of metals in leachate from soil contacted with an acetic acid solution (TCLP) [60] or a dilute solution of sulfuric and nitric acid (SPLP). In this case, metal contamination is expressed in $mgL^{-1}$ of the leachable metal. Other types of leaching tests have been proposed including sequential extraction procedures [61, 62] and extraction of acid volatile sulfide [63]. Sequential procedures contact the solid with a series of extractant solutions that are designed to dissolve different fractions of the associated metal. These tests may provide insight into the different forms of metal contamination present. Contaminant concentrations can be measured directly in metals-contaminated water. These concentrations are most commonly expressed as total dissolved metals in mass concentrations ($mg\,L^{-1}$ or $gL^{-1}$) or in molar concentrations ($mol\,L^{-1}$). In dilute solutions, a mg $L^{-1}$ is equivalent to one part per million (ppm), and a g$L^{-1}$ is equivalent to one part per billion (ppb).

Riley et al. [64] and NJDEP [65] have reported soil concentration ranges and regulatory guidelines for some heavy metals (Table 1). In Nigeria, in the interim period, whilst suitable parameters are being developed, the Department of Petroleum Resources [60] has recommended guidelines on remediation of contaminated land based on two parameters intervention values and target values (Table 2).

**Table 1:** Soil concentration ranges and regulatory guidelines for some heavy metals

| Metal | Soil concentration range† | Regulatory limits‡ |
|---|---|---|
| | (mg $kg^{-1}$) | (mg $kg^{-1}$) |
| Pb | 1.00–69 000 | 600 |
| Cd | 0.10–345 | 100 |
| Cr | 0.05–3 950 | 100 |
| Hg | <0.01–1 800 | 270 |
| Zn | 150–5 000 | 1 500 |

[64]; ‡Nonresidential direct contact soil clean-up criteria [65].

**Table 2:** Target and intervention values for some metals for a standard soil [60]

| Metal | Target value | Intervention value |
|---|---|---|
| | (mg $kg^{-1}$) | (mg $kg^{-1}$) |
| Ni | 140.00 | 720.00 |
| Cu | 0.30 | 10.00 |
| Zn | — | — |
| Cd | 100.00 | 380.00 |
| Pb | 35.00 | 210.00 |
| As | 200 | 625 |
| Cr | 20 | 240 |
| Hg | 85 | 530 |

The intervention values indicate the quality for which the functionality of soil for human, animal, and plant life are, or threatened with being seriously impaired. Concentrations in excess of the intervention values correspond to serious contamination. Target values indicate the soil quality required for sustainability or expressed in terms of remedial policy, the soil quality required for the full restoration of the soil's functionality for human, animal, and plant life. The target values therefore indicate the soil quality levels ultimately aimed at.

# REMEDIATION OF HEAVY METAL-CONTAMINATED SOILS

The overall objective of any soil remediation approach is to create a final solution that is protective of human health and the environment [66]. Remediation is generally subject to an array of regulatory requirements and can also be based on assessments of human health and ecological risks where no legislated standards exist or where standards are advisory. The regulatory authorities will normally accept remediation strategies that centre on reducing metal bioavailability only if reduced bioavailability is equated with reduced risk, and if the bioavailability reductions are demonstrated to be long term [66]. For heavy metal contaminated soils, the physical and chemical form of the heavy metal contaminant in soil strongly influences the selection of the appropriate remediation treatment approach. Information about the physical characteristics of the site and the type and level of contamination at the site must be obtained to enable accurate assessment of site contamination and remedial alternatives. The contamination in the soil should be characterized to establish the type, amount, and distribution of heavy metals in the soil. Once the site has been characterized, the desired level of each metal in soil must be determined. This is done by comparison of observed heavy metal concentrations with soil quality standards for a particular regulatory domain, or by performance of a site-specific risk assessment. Remediation goals for heavy metals may be set as total metal concentration or as leachable metal in soil, or as some combination of these.

Several technologies exist for the remediation of metal-contaminated soil. Gupta et al. [67] have classified remediation technologies of contaminated soils into three categories of hazard-alleviating measures: (i) gentlein situ remediation, (ii) in situ harsh soil restrictive measures, and (iii) in situ or ex situ harsh soil destructive measures. The goal of the last two harsh alleviating measures is to avert hazards either to man, plant, or animal while the main goal of gentle in situ remediation is to restore the malfunctionality of soil (soil fertility), which allows a safe use of the soil. At present, a variety of approaches have been suggested for remediating contaminated soils. USEPA [68] has broadly classified remediation technologies for contaminated soils into (i) source control and (ii) containment remedies. Source control involves in situ and ex situ treatment technologies for sources of contamination. In situ or in place means that the contaminated soil is treated in its original place; unmoved, unexcavated; remaining at the site or in the subsurface. In situ treatment technologies treat or remove the contaminant from soil without excavation or removal of the soil. Ex situ means that the contaminated soil is moved, excavated, or removed from the site or subsurface. Implementation of ex situremedies requires excavation or removal of the contaminated soil. Containment remedies involve the construction of vertical engineered barriers (VEB), caps, and liners used to prevent the migration of contaminants.

Another classification places remediation technologies for heavy metal-contaminated soils under five categories of general approaches to remediation (Table 3): isolation, immobilization, toxicity reduction, physical separation, and extraction [3]. In practice, it may be more convenient to employ a hybrid of two or more of these approaches for more cost effectiveness. The key factors that may influence the applicability and selection of any of the available remediation technologies are: (i) cost, (ii) long-term effectiveness/permanence, (iii) commercial availability, (iv) general acceptance, (v) applicability to high metal concentrations, (vi) applicability to mixed wastes (heavy metals and organics), (vii) toxicity reduction, (viii) mobility reduction, and (ix) volume reduction. The present paper focuses on soil washing, phytoremediation, and immobilization techniques since they are among the best demonstrated available technologies (BDATs) for heavy metal-contaminated sites.

**Table 3:** Technologies for remediation of heavy metal-contaminated soils

| Category | Remediation technologies |
|---|---|
| Isolation | (i) Capping (ii) subsurface barriers. |
| Immobilization | (i) Solidification/stabilization (ii) vitrification (iii) chemical treatment. |
| Toxicity and/or mobility reduction | (i) Chemical treatment (ii) permeable treatment walls (iii) biological treatment bioaccumulation, phytoremediation (phytoextraction, phytostabilization, and rhizofiltration), bioleaching, biochemical processes. |
| Physical separation | |
| Extraction | (i) Soil washing, pyrometallurgical extraction, in situ soil flushing, and electrokinetic treatment. |

## Immobilization Techniques

Ex situ and in situ immobilization techniques are practical approaches to remediation of metal-contaminated soils. The ex situ technique is applied in areas where highly contaminated soil must be removed from its place of origin, and its storage is connected with a high ecological risk (e.g., in the case of radio nuclides). The method's advantages are: (i) fast and easy applicability and (ii) relatively low costs of investment and operation. The method's disadvantages include (i) high invasivity to the environment, (ii) generation of a significant amount of solid wastes (twice as large as volume after processing), (iii) the byproduct must be stored on a special landfill site, (iv) in the case of changing of the physicochemical condition in the side product or its surroundings, there is serious danger of the release of additional contaminants to the environment, and (v) permanent control of the stored wastes is required. In the in situ technique, the fixing agents amendments are applied on the unexcavated soil. The technique's advantages are (i) its low invasivity, (ii) simplicity and rapidity, (iii) relatively inexpensive, and (iv) small amount of wastes are produced, (v) high public acceptability, (vi) covers a broad spectrum of inorganic pollutants. The disadvantages of in situ immobilization are (i) its only a temporary solution (contaminants are still in the environment), (ii) the activation of pollutants may occur when soil physicochemical properties change, (iii) the reclamation process is applied only to the surface layer of soil (30–50 cm), and (iv) permanent monitoring is necessary [66, 69].

Immobilization technology often uses organic and inorganic amendment to accelerate the attenuation of metal mobility and toxicity in soils. The primary role of immobilizing amendments is to alter the original soil metals to more geochemically stable phases via sorption, precipitation, and complexation processes [70]. The mostly applied amendments include clay, cement, zeolites, minerals, phosphates, organic composts, and microbes [3, 71]. Recent studies have indicated the potential of low-cost industrial residues such as red mud [72, 73] andtermitaria [74] in immobilization of heavy metals in contaminated soils. Due to the complexity of soil matrix and the limitations of current analytical techniques, the exact immobilization mechanisms have not been clarified, which could include precipitation, chemical adsorption and ion exchange, surface precipitation, formation of stable complexes with organic ligands, and redox reaction [75]. Most immobilization technologies can be performed ex situ or in situ. In situ processes are preferred due to the lower labour and energy requirements, but implementation of in situ will depend on specific site conditions.

## Solidification/Stabilization (S/S)

Solidification involves the addition of binding agents to a contaminated material to impart physical/dimensional stability to contain contaminants in a solid product and reduce access by external agents through a combination of chemical reaction, encapsulation, and reduced permeability/surface area. Stabilization (also referred to as fixation) involves the addition of reagents to the contaminated soil to produce more chemically stable constituents. Conventional S/S is an established remediation technology for contaminated soils and treatment technology for hazardous wastes in many countries in the world [76].

The general approach for solidification/stabilization treatment processes involves mixing or injecting treatment agents to the contaminated soils. Inorganic binders (Table 4), such as clay (bentonite and kaolinite), cement, fly ash, blast furnace slag, calcium carbonate, Fe/Mn oxides, charcoal, zeolite [9, 77], and organic stabilizers (Table5) such as bitumen, composts, and manures [78], or a combination of organic-inorganic amendments may be used. The dominant mechanism by which metals are immobilized is by precipitation of hydroxides within the solid matrix [79, 80]. Solidification/stabilization technologies are not useful for some forms of metal contamination, such as species that exist as oxyanions (e.g., $Cr_2O_7^{2-}$, $AsO_3^{-}$) or metals that do not have low-solubility hydroxides (e.g., Hg). Solidification/stabilization may not be applicable at sites containing wastes that include organic forms of contamination, especially if volatile organics are present. Mixing and heating associated with binder hydration may release organic vapors. Pretreatment, such as air stripping or incineration, may be used to remove the organics and prepare the waste for metal stabilization/ solidification [39]. The application of S/S technologies will also be affected by the chemical composition of the contaminated matrix, the amount of water present, and the ambient temperature. These factors can interfere with the solidification/stabilization process by inhibiting bonding of the waste to the binding material, retarding the setting of the mixtures, decreasing the stability of the matrix, or reducing the strength of the solidified area [81].

**Table 4:** Organic amendments for heavy metal immobilization [82]

| Material | Heavy metal immobilized |
|---|---|
| Bark saw dust (from timber industry) | Cd, Pb, Hg, Cu |
| Xylogen (from paper mill wastewater) | Zn, Pb, Hg |
| Chitosan (from crab meat canning industry) | Cd, Cr, Hg |
| Bagasse (from sugar cane) | Pb |
| Poultry manure (from poultry farm) | Cu, Pb, Zn, Cd |
| Cattle manure (from cattle farm) | Cd |
| Rice hulls (from rice processing) | Cd, Cr, Pb |
| Sewage sludge | Cd |
| Leaves | Cr, Cd |
| Straw | Cd, Cr, Pb |
| | |

**Table 5:** Inorganic amendments for heavy metal immobilization [82]

| Material | Heavy metal immobilized |
|---|---|
| Lime (from lime factory) | Cd, Cu, Ni, Pb, Zn, |
| Phosphate salt (from fertilizer plant) | Pb, Zn, Cu, Cd |

| | |
|---|---|
| Hydroxyapatite (from phosphorite) | Zn, Pb, Cu, Cd |
| Fly ash (from thermal power plant) | Cd, Pb, Cu, Zn, Cr |
| Slag (from thermal power plant) | Cd, Pb, Zn, Cr |
| Ca-montmorillonite (mineral) | Zn, Pb |
| Portland cement (from cement plant) | Cr, Cu, Zn, Pb |
| Bentonite | Pb |

Cement-based binders and stabilizers are common materials used for implementation of S/S technologies [83]. Portland cement, a mixture of Ca silicates, aluminates, aluminoferrites, and sulfates, is an important cement-based material. Pozzolanic materials, which consist of small spherical particles formed by coal combustion (such as fly ash) and in lime and cement kilns, are also commonly used for S/S. Pozzolans exhibit cement-like properties, especially if the silica content is high. Portland cement and pozzolans can be used alone or together to obtain optimal properties for a particular site [84]. Organic binders may also be used to treat metals through polymer microencapsulation. This process uses organic materials such as bitumen, polyethylene, paraffins, waxes, and other polyolefins as thermoplastic or thermosetting resins. For polymer encapsulation, the organic materials are heated and mixed with the contaminated matrix at elevated temperatures (120° to 200°C). The organic materials polymerize and agglomerate the waste, and the waste matrix is encapsulated [84]. Organics are volatilized and collected, and the treated material is extruded for disposal or possible reuse (e.g., as paving material) [39]. The contaminated material may require pretreatment to separate rocks and debris and dry the feed material. Polymer encapsulation requires more energy and more complex equipment than cement-based S/S operations. Bitumen (asphalt) is the cheapest and most common thermoplastic binder [84]. Solidification/stabilization is achieved by mixing the contaminated material with appropriate amounts of binder/stabilizer and water. The mixture sets and cures to form a solidified matrix and contain the waste. The cure time and pour characteristics of the mixture and the final properties of the hardened cement depend upon the composition (amount of cement, pozzolan, and water) of the binder/stabilizer.

Ex situ S/S can be easily applied to excavated soils because methods are available to provide the vigorous mixing needed to combine the binder/stabilizer with the contaminated material. Pretreatment of the waste may be necessary to screen and crush large rocks and debris.

Mixing can be performed via in-drum, in-plant, or area-mixing processes. In-drum mixing may be preferred for treatment of small volumes of waste or for toxic wastes. In-plant processes utilize rotary drum mixers for batch processes or pug mill mixers for continuous treatment. Larger volumes of waste may be excavated and moved to a contained area for area mixing. This process involves layering the contaminated material with the stabilizer/binder, and subsequent mixing with a backhoe or similar equipment. Mobile and fixed treatment plants are available for ex situ S/S treatment. Smaller pilot-scale plants can treat up to 100 tons of contaminated soil per day while larger portable plants typically process 500 to over 1000 tons per day [39]. Stabilization/stabilization techniques are available to provide mixing of the binder/stabilizer with the contaminated soil in situ. In situ S/S is less labor and energy intensive than ex situ process that require excavation, transport, and disposal of the treated material. In situ S/S is also preferred if volatile or semivolatile organics are present because excavation would expose these contaminants to the air [85]. However, the presence of bedrock, large boulders cohesive soils, oily sands, and clays may preclude the application of in situ S/S at some sites. It is also more difficult to provide uniform and complete mixing through in situ processes. Mixing of the binder and contaminated matrix may be achieved using in-place mixing, vertical auger mixing, or injection grouting. In-place mixing is similar to ex situ area mixing except that the soil is not excavated prior to treatment. The in situ process is useful for treating surface or shallow contamination and involves spreading and mixing the binders with the waste using conventional excavation equipment such as draglines, backhoes, or clamshell buckets. Vertical auger mixing uses a system of augers to inject and mix the binding reagents with the waste. Larger (6–12 ft diameter) augers are used for shallow (10–40 ft) drilling and can treat 500–1000 cubic yards per day [86, 87]. Deep stabilization/solidification (up to 150 ft) can be achieved by using ganged augers (up to 3 ft in diameter each) that can treat 150–400 cubic yards per day. Finally injection grouting may be performed to inject the binder containing suspended or dissolved reagents into the treatment area under pressure. The binder permeates the surrounding soil and cures in place [39].

## *Vitrification*

The mobility of metal contaminants can be decreased by high-temperature treatment of the contaminated area that results in the formation of vitreous material, usually an oxide solid. During this process, the increased temperature may also volatilize and/or destroy organic contaminants or volatile metal species (such as Hg) that must be collected for treatment or disposal. Most soils can be treated by vitrification, and a wide variety of inorganic and organic contaminants can be targeted. Vitrification may be performed ex situ or in situ althoughin situ processes are preferred due to the lower energy requirements and cost [88]. Typical stages in ex situvitrification processes may include excavation, pretreatment, mixing, feeding, melting and vitrification, off-gas collection and treatment, and forming or casting of the melted product. The energy requirement for melting is the primary factor influencing the cost of ex situ vitrification. Different sources of energy can be used for this purpose, depending on local energy costs. Process heat losses and water content of the feed should be controlled in order to minimize energy requirements. Vitrified material with certain characteristics may be obtained by using additives such as sand, clay, and/or native soil. The vitrified waste may be recycled and used as clean fill, aggregate, or other reusable materials [39]. In situ vitrification (ISV) involves passing electric current through the soil using an array of electrodes inserted vertically into the contaminated region. Each setting of four electrodes is referred to as a melt. If the soil is too dry, it may not provide sufficient conductance, and a trench containing flaked graphite and glass frit (ground glass particles) must be placed between the electrodes to provide an initial flow path for the current. Resistance heating in the starter path melts the soil. The melt grows outward and down as the molten soil usually provides additional conductance for the current. A single melt can treat up to 1000 tons of contaminated soil to depths of 20 feet, at a typical treatment rate of 3 to 6 tons per hour. Larger areas are treated by fusing together multiple individual vitrification zones. The main requirement for in situ vitrification is the ability of the soil melt to carry current and solidify as it cools. If the alkali content (as $Na_2O$ and $K_2O$) of the soil is too high (1.4 wt%), the molten soil may not provide enough conductance to carry the current [89].

Vitrification is not a classical immobilization technique. The advantages include (i) easily applied for reclamation of heavily contaminated soils (Pb, Cd, Cr, asbestos, and materials containing asbestos), (ii) in the course of applying this method qualification of wastes (from hazardous to neutral) could be changed.

## Assessment of Efficiency and Capacity of Immobilization

The efficiency (*E*) and capacity (*P*) of different additives for immobilization and field applications can be evaluated using the expressions

$$E(\%) = \frac{M_o - M_e}{M_0} \times 100,$$

$$p = \frac{(M_o - M_e)V}{m}, \qquad (2)$$

Where $E$ = efficiency of immobilization agent; $P$ = capacity of immobilization agent; $Me$ – equilibrium extractable concentration of single metal in the immobilized soil (mg L–1); $Mo$ = initial extractable concentration of single metal in preimmobilized soil (mg L–1); $V$ = volume of metal salt solution (mg L–1); $m$ = weight of immobilization agent (g) [90]. High values of $E$ and $P$ represent the perfect efficiency and capacity of an additive that can be used in field studies of metal immobilization. After screening out the best efficient additive, another experiment could be conducted to determine the best ratio (soil/additive) for the field-fixing treatment. After the fixing treatment of contaminated soils, a lot of methods including biological and physiochemical experiments could be used to assess the remediation efficiency. Environmental risk could also be estimated after confirming the immobilized efficiency and possible release [89].

## Soil Washing

Soil washing is essentially a volume reduction/waste minimization treatment process. It is done on the excavated (physically removed) soil (ex situ) or on-site (in situ). Soil washing as discussed in this review refers to ex situ techniques that employ physical and/or chemical procedures to extract metal contaminants from soils. During soil washing, (i) those soil particles which host the majority of the contamination are separated from the bulk soil fractions (physical separation), (ii) contaminants are removed from the soil by aqueous chemicals and recovered from solution on a solid substrate (chemical extraction), or (iii) a combination of both [91]. In all cases, the separated contaminants then go to hazardous waste landfill (or occasionally are further treated by chemical, thermal, or biological processes). By removing the majority of the contamination from the soil, the bulk fraction that remains can be (i) recycled on the site being remediated as relatively inert backfill, (ii) used on another site as fill, or (iii) disposed of relatively cheaply as nonhazardous material.

Ex situ soil washing is particularly frequently used in soil remediation because it (i) completely removes the contaminants and hence ensures the rapid cleanup of a contaminated site [92], (ii) meets specific criteria, (iii) reduces or eliminates long-term liability, (iv) may be the most cost-effective solution, and (v) may produce recyclable material or energy [93]. The disadvantages include the fact that the contaminants are simply moved to a different place, where they must be monitored, the risk of spreading contaminated soil and dust particles during removal and transport of contaminated soil, and the relatively high cost. Excavation can be the most expensive option when large amounts of soil must be removed, or disposal as hazardous or toxic waste is required.

Acid and chelator soil washing are the two most prevalent removal methods [94]. Soil washing currently involves soil flushing an in situ process in which the washing solution is forced through the in-place soil matrix, ex situ extraction of heavy metals from the soil slurry in reactors, and soil heap leaching. Another heavy metal removal technology is electroremediation, which mostly involves electrokinetic movement of charged particles suspended in the soil solution, initiated by an electric gradient [35]. The metals can be removed by precipitation at the electrodes. Removal of the majority of the contaminants from the soil does not mean that the contaminant-depleted bulk is totally contaminant free. Thus, for soil washing to be successful, the level of contamination in the treated bulk must be below a site-specific action limit (e.g., based on risk assessment). Cost effectiveness with soil washing is achieved by offsetting processing costs against the ability to significantly reduce the amount of material requiring costly disposal at a hazardous waste landfill [95].

Typically the cleaned fractions from the soil washing process should be >70–80% of the original mass of the soil, but, where the contaminants have a very high associated disposal cost, and/or where transport distances to the nearest hazardous waste landfill are substantial, a 50% reduction might still be cost effective. There is also a generally held opinion that soil washing based on physical separation processes is only cost effective for sandy and granular soils where the clay and silt content (particles less than 0.063 mm) is less than 30–35% of the soil. Soil washing by chemical dissolution of the contaminants is not constrained by the proportion of clay as this fraction can also be leached by the chemical agent. However, clay-rich soils pose other problems such as difficulties with materials handling and solid-liquid separation [96]. Full-scale soil washing plants exist as fixed centralized treatment centres, or as mobile/transportable units. With fixed centralized facilities, contaminated soil is brought to the plant, whereas with mobile/transportable facilities, the plant is transported to a contaminated site, and soil is processed on the site. Where mobile/transportable plant is used, the cost of mobilization and demobilization can be significant. However, where large volumes of soil are to be treated, this cost can be more than offset by reusing clean material on the site (therefore avoiding the cost of transport to an off-site centralized treatment facility, and avoiding the cost of importing clean fill).

## Principles of Soil Washing

Soil washing is a volume reduction/waste minimization treatment technology based on physical and/or chemical processes. With physical soil washing, differences between particle grain size, settling velocity, specific gravity, surface chemical behaviour, and rarely magnetic properties are used to separate those particles whichhost the majority of the contamination from the bulk which are contaminant-depleted. The equipment used is standard mineral processing equipment, which is more generally used in the mining industry [91]. Mineral processing techniques as applied to soil remediation have been reviewed in literature [97].

With chemical soil washing, soil particles are cleaned by selectively transferring the contaminants on the soil into solution. Since heavy metals are sparingly soluble and occur predominantly in a sorbed state, washing the soils with water alone would be expected to remove too low an amount of cations in the leachates, chemical agents have to be added to the washing water [98]. This is achieved by mixing the soil with aqueous solutions of acids, alkalis, complexants, other solvents, and surfactants. The resulting cleaned particles are then separated from the resulting aqueous solution. This solution is then treated to remove the contaminants (e.g., by sorption on activated carbon or ion exchange) [91, 95].

The effectiveness of washing is closely related to the ability of the extracting solution to dissolve the metal contaminants in soils. However, the strong bonds between the soil and metals make the cleaning process difficult [99]. Therefore, only extractants capable of dissolving large quantities of metals would be suitable for cleaning purposes. The realization that the goal of soil remediation is to remove the metal and preserve the natural soil properties limits the choice of extractants that can be used in the cleaning process [100].

## Chemical Extractants for Soil Washing

Owing to the different nature of heavy metals, extracting solutions that can optimally remove them must be carefully sought during soil washing. Several classes of chemicals used for soil washing include surfactants, cosolvents, cyclodextrins, chelating agents, and organic acids [101–106]. All these soil washing extractants have been developed on a case-by-case basis depending on the contaminant type at a particular site. A few studies have indicated that the solubilization/exchange/extraction of heavy metals by washing solutions differs considerably for different soil types. Strong acids attack and degrade the soil crystalline structure at extended contact times. For less damaging washes, organic acids and chelating agents are often suggested as alternatives to straight mineral acid use [107].

Natural, low-molecular-weight organic acids (LMWOAs) including oxalic, citric, formic, acetic, malic, succinic, malonic, maleic, lactic, aconitic, and fumaric acids are natural products of root exudates, microbial secretions, and plant and animal residue decomposition in soils [108]. Thus metal dissolution by organic acids is likely to be more representative of a mobile metal fraction that is available to biota [109]. The chelating organic acids are able to dislodge the exchangeable, carbonate, and reducible fractions of heavy metals by washing procedures [94]. Although many chelating compounds including citric acid [108], tartaric acid [110], and EDTA [94, 100, 111] for mobilizing heavy metals have been evaluated, there remain uncertainties as to the optimal choice for full-scale application. The identification and quantification of coexisting solid metal species in the soil before and after treatment are essential to design and assess the efficiency of soil-washing technology [4]. A recent study [112] showed that changes in Ni, Cu, Zn, Cd, and Pb speciation and uptake by maize in a sandy loam before and after washing with three chelating organic acids indicated that EDTA and citric acid appeared to offer greater potentials as chelating agents for remediating the permeable soil. Tartaric acid was, however, recommended in events of moderate contamination.

The use of soil washing to remediate contaminated fine-grained soils that contained more than 30% fines fraction has been reported by several workers [113–115]. Khodadoust et al. [59, 116] have also studied the removal of various metals (Pb, Ni, and Zn) from field and clay (kaolin) soil samples using a broad spectrum of extractants (chelating agents and organic acids). Chen and Hong [117] reported on the chelating extraction of Pb and Cu from an authentic contaminated soil using derivatives of iminodiacetic acid and L-cyestein. Wuana et al. [118] investigated the removal of Pb and Cu from kaolin and bulk clay soils using two mineral acids ($HCl$ and $H_2SO_4$) and chelating agents (EDTA and oxalic acid). The use of chelating organic acids—citric acid, tartaric acid and EDTA in the simultaneous removal of Ni, Cu, Zn, Cd, and Pb from an experimentally contaminated sandy loam was carried out by Wuana et al. [112]. These studies furnished valuable information on the distribution of heavy metals in the soils and their removal using various extracting solutions.

## Phytoremediation

Phytoremediation, also called green remediation, botanoremediation, agroremediation, or vegetative remediation, can be defined as an in situ remediation strategy that uses vegetation and associated microbiota, soil amendments, and agronomic techniques to remove, contain, or render environmental contaminants harmless [119, 120]. The idea of using metal-accumulating plants to remove heavy metals and other compounds was first introduced in 1983, but the concept has actually been implemented for the past 300 years on wastewater discharges [121, 122]. Plants may break down or degrade organic pollutants or remove and stabilize metal contaminants. The methods used to phytoremediate metal contaminants are slightly different from those used to remediate sites polluted with organic contaminants. As it is a relatively new technology, phytoremediation is still mostly in its testing stages and as such has not been used in many places as a full-scale application. However, it has been tested successfully in many places around the world for many different contaminants. Phytoremediation is energy efficient, aesthetically pleasing method of remediating sites with low-to-moderate levels of contamination, and it can be used in conjunction with other more traditional remedial methods as a finishing step to the remedial process.

The advantages of phytoremediation compared with classical remediation are that (i) it is more economically viable using the same tools and supplies as agriculture, (ii) it is less disruptive to the environment and does not involve waiting for new plant communities to recolonize the site, (iii) disposal sites are not needed, (iv) it is more likely to be accepted by the public as it is more aesthetically pleasing then traditional methods, (v) it avoids excavation and transport of polluted media thus reducing the risk of spreading the contamination, and (vi) it has the potential to treat sites polluted with more than one type of pollutant. The disadvantages are as follow (i) it is dependant on the growing conditions required by the plant (i.e., climate, geology, altitude, and temperature), (ii) large-scale operations require access to agricultural equipment and knowledge, (iii) success is dependant on the tolerance of the plant to the pollutant, (iv) contaminants collected in senescing tissues may be released back into the environment in autumn, (v) contaminants may be collected in woody tissues used as fuel, (vi) time taken to remediate sites far exceeds that of other technologies, (vii) contaminant solubility may be increased leading to greater environmental damage and the possibility of leaching. Potentially useful phytoremediation technologies for remediation of heavy metal-contaminated soils include phytoextraction (phytoaccumulation), phytostabilization, and phytofiltration [123].

## Phytoextraction (Phytoaccumulation)

Phytoextraction is the name given to the process where plant roots uptake metal contaminants from the soil and translocate them to their above soil tissues. A plant used for phytoremediation needs to be heavy-metal tolerant, grow rapidly with a high biomass yield per hectare, have high metal-accumulating ability in the foliar parts, have a profuse root system, and a high bioaccumulation factor [21, 124]. Phytoextraction is, no doubt, a publicly appealing (green) remediation technology [125]. Two approaches have been proposed for phytoextraction of heavy metals, namely, continuous or natural phytoextraction and chemically enhanced phytoextraction [126, 127].

## Continuous or Natural Phytoextraction

Continuous phytoextraction is based on the use of natural hyperaccumulator plants with exceptional metal-accumulating capacity. Hyperaccumulators are species capable of accumulating metals at levels 100 fold greater than those typically measured in shoots of the common nonaccumulator plants. Thus, a hyperaccumulator plant will concentrate more than 10 mg $kg^{-1}$ Hg, 100 mg $kg^{-1}$ Cd, 1000 mg $kg^{-1}$ Co, Cr, Cu, and Pb; 10 000 mg $kg^{-1}$Zn and Ni [128, 129]. Hyperaccumulator plant species are used on metalliferous sites due to their tolerance of relatively high levels of pollution. Approximately 400 plant species from at least 45 plant families have been so far, reported to hyperaccumulate metals [22, 127]; some of the families are Brassicaceae, Fabaceae, Euphorbiaceae, Asterraceae, Lamiaceae, and Scrophulariaceae [130, 131]. Crops like alpine pennycress (Thlaspi caerulescens), Ipomea alpine, Haumaniastrum robertii, Astragalus racemosus, Sebertia acuminate have very high bioaccumulation potential for Cd/Zn, Cu, Co, Se, and Ni, respectively [22]. Willow (Salix viminalis L.), Indian mustard (Brassica juncea L.), corn (Zea mays L.), and sunflower (Helianthus annuus L.) have reportedly shown high uptake and tolerance to heavy metals [132]. A list of some plant hyperaccumulators are given in Table 6. A number of processes are involved during phytoextraction of metals from soil: (i) a metal fraction is sorbed at root surface, (ii) bioavailable metal moves across cellular membrane into root cells, (iii) a fraction of the metal absorbed into roots is immobilized in the vacuole, (iv) intracellular mobile metal crosses cellular membranes into root vascular tissue (xylem), and (v) metal is translocated from the root to aerial tissues (stems and leaves) [22]. Once inside the plant, most metals are too insoluble to move freely in the vascular system so they usually form carbonate, sulphate, or phosphate precipitate immobilizing them in apoplastic (extracellular) and symplastic (intracellular) compartments [46]. Hyperaccumulators have several beneficial characteristics but may tend to be slow growing and produce low biomass, and years or decades are needed to clean up contaminated sites. To overcome these shortfalls, chemically enhanced phytoextraction has been developed. The approach makes use of high biomass crops that are induced to take up large amounts of metals when their mobility in soil is enhanced by chemical treatment with chelating organic acids [133].

**Table 6:** Some metal hyperaccumulating plants [21]

| Plant | Metal | Concentration (mg $kg^{-1}$) |
|---|---|---|
| Dicotyledons | | |
| Cystus ladanifer | Cd | 309 |
| | Co | 2 667 |
| | Cr | 2 667 |
| | Ni | 4 164 |
| | Zn | 7 695 |
| Thlaspi caerulescens | Cd | 10 000–15 000 |
| | Zn | 10 000–15 000 |
| Arabidopsis halleri | Cd | 5 900–31 000 |
| Alyssum sp. | Ni | 4 200–24 400 |
| Brassica junica | Pb | 10 000–15 000 |
| | Zn | 2 600 |
| Betula | Zn | 528 |
| Grasses | | |
| Vetiveria zizaniodes | Zn | 0.03 |
| Paspalum notatum | | |
| Stenotaphrum secundatum | | |
| Pennisetum glaucum | | |

## *Chelate-Assisted (Induced) Phytoextraction*

For more than 10 years, chelant-enhanced phytoextraction of metals from contaminated soils have received much attention as a cost-effective alternative to conventional techniques of enhanced soil remediation [133,134]. When the chelating agent is applied to the soil, metal-chelant complexes are formed and taken up by the plant, mostly through a passive apoplastic pathway [133]. Unless the metal ion is transported as a noncationic chelate, apoplastic transport is further limited by the high cation exchange capacity of cell walls [46]. Chelators have been isolated from plants that are strongly involved in the uptake of heavy metals and their detoxification. The chelating agent EDTA has become one of the most tested mobilizing amendments for less mobile/ available metals such as Pb [135, 136]. Chelators have been isolated from plants that are strongly involved in the uptake of heavy metals and their detoxification. The addition of EDTA to a Pb-contaminated soil (total soil Pb 2500 mg $kg^{-1}$) increased shoot lead concentration of Zea mays L. (corn) and Pisun sativum (pea) from less than 500 mg $kg^{-1}$ to more than 10,000 mg $kg^{-1}$. Enhanced accumulation of metals by plant species with EDTA treatment is attributed to many factors working either singly or in combination. These factors include (i) an increase in the concentration of available metals, (ii) enhanced metal-EDTA complex movement to roots, (iii) less binding of metal-EDTA complexes with the negatively charged cell wall constituents, (iv) damage to physiological barriers in roots either due to greater concentration of metals or EDTA or metal-EDTA complexes, and (v) increased mobility of metals within the plant body when complexed with EDTA compared to free-metal ions facilitating the translocation of metals from roots to shoots [134, 137]. For the chelates tested, the order of effectiveness in increasing Pb desorption from the soil was EDTA > hydroxyethylethylene-diaminetriacetic acid (HEDTA) > diethylenetriaminepentaacetic acid (DTPA) > ethylenediamine di(o-hyroxyphenylacetic acid) EDDHA [135]. Vassil et al. [138] reported that Brassica juncea exposed to Pb and EDTA in hydroponic solution was able to accumulate up to 55 mM $kg^{-1}$ Pb in dry shoot tissue (1.1% w/w). This represents a 75-fold concentration of lead in shoot over that in solution. A 0.25 mM threshold concentration of EDTA was required to stimulate this dramatic accumulation of both lead and EDTA in shoots. Since EDTA has been associated with high toxicity and persistence in the environment, several other alternatives have been proposed. Of all those, EDDS ([S,S]-ethylenediamine disuccinate) has been introduced as a promising and environmentally friendlier mobilizing agent, especially for Cu and Zn [135, 139, 140]. Once the plants have grown and absorbed the metal pollutants, they are harvested and disposed of safely. This process is repeated several times to reduce contamination to acceptable levels. Interestingly, in the last few years, the possibility of planting metal hyperaccumulator crops over a low-grade ore body or mineralized soil, and then harvesting and incinerating the biomass to produce a commercial bio-ore has been proposed [141] though this is usually reserved for use with precious metals. This process calledphytomining offers the possibility of exploiting ore bodies that are otherwise uneconomic to mine, and its effect on the environment is minimal when compared with erosion caused by opencast mining [123, 141].

## Assessing the Efficiency of Phytoextraction

Depending on heavy metal concentration in the contaminated soil and the target values sought for in the remediated soil, phytoextraction may involve repeated cropping of the plant until the metal concentration drops to acceptable levels. The ability of the plant to account for the decrease in soil metal concentrations as a function of metal uptake and biomass production plays an important role in achieving regulatory acceptance. Theoretically, metal removal can be accounted for by determining metal concentration in the plant, multiplied by the reduction in soil metal concentrations [127]. It should, however, be borne in mind that this approach may be challenged by a number of factors working together during field applications. Practically, the bioaccumulation factor, $f$, amount of metal extracted, $M$ (mg/kg plant) and phytoremediation time, $tp$ (year) [142] can be used to evaluate the plant's phytoextraction efficiency and calculated according to equation (3) [143] by assuming that the plant can be cropped $n$ times each year and metal pollution occurs only in the active rooting zone, that is, top soil layer (0–20 cm) and still assuming a soil bulk density of 1.3 t/m$^3$, giving a total soil mass of 2600 t/ha.

$$f = \frac{Metal\ concentration\ in\ plant\ shoot}{Metal\ concentration\ in\ soil},$$

$$M(mg\ /\ kg\ plant) = Metal\ concentration\ in\ planttissue \times Biomass,$$

$$t_p(year) = \frac{Metal\ concentration\ in\ soi\ \ln eededtodecrease \times Soil\ mass}{Metal\ concentration\ in\ plant\ shoot \times Plant\ shoot\ biomass \times n}. \quad (3)$$

## Prospects of Phytoextraction

One of the key aspects of the acceptance of phytoextraction pertains to its performance, ultimate utilization of byproducts, and its overall economic viability. Commercialization of phytoextraction has been challenged by the expectation that site remediation should be achieved in a time comparable to other clean-up technologies [123]. Genetic engineering has a great role to play in supplementing the list of plants available for phytoremediation by the use of engineering tools to insert into plants those genes that will enable the plant to metabolize a particular pollutant [144]. A major goal of plant genetic engineering is to enhance the ability of plants to metabolize many of the compounds that are of environmental concern. Currently, some laboratories are using traditional breeding techniques, others are creating protoplast-fusion hybrids, and still others are looking at the direct insertion of novel genes to enhance the metabolic capabilities of plants [144]. On the whole, phytoextraction appears a very promising technology for the removal of metal pollutants from the environment and is at present approaching commercialization.

## Possible Utilization of Biomass after Phytoextraction

A serious challenge for the commercialization of phytoextraction has been the disposal of contaminated plant biomass especially in the case of repeated cropping where large tonnages of biomass may be produced. The biomass has to be stored, disposed of or utilized in an appropriate manner so as not to pose any environmental risk. The major constituents of biomass material are lignin, hemicellulose, cellulose, minerals, and ash. It possesses high moisture and volatile matter, low bulk density, and calorific value [127]. Biomass is solar energy fixed in plants in form of carbon, hydrogen, and oxygen (oxygenated hydrocarbons) with a possible general chemical formula $CH_{1.44}O_{0.66}$. Controlled combustion and gasification of biomass can yield a mixture of producer gas and/or pyro-gas which leads to the generation of thermal and electrical energy [145]. Composting and compacting can be employed as volume reduction approaches to biomass reuse [146]. Ashing of biomass can produce bio-ores especially after the phytomining of precious metals. Heavy metals such as Co, Cu, Fe, Mn, Mo, Ni, and Zn are plant essential metals, and most plants have the ability to accumulate them [147]. The high concentrations of these metals in the harvested biomass can be "diluted" to acceptable concentrations by combining the biomass with clean biomass in formulations of fertilizer and fodder.

## *Phytostabilization*

Phytostabilization, also referred to as in-place inactivation, is primarily concerned with the use of certain plants to immobilize soil sediment and sludges [148]. Contaminant are absorbed and accumulated by roots, adsorbed onto the roots, or precipitated in the rhizosphere. This reduces or even prevents the mobility of the contaminants preventing migration into the groundwater or air and also reduces the bioavailability of the contaminant thus preventing spread through the food chain. Plants for use in phytostabilization should be able to (i) decrease the amount of water percolating through the soil matrix, which may result in the formation of a hazardous leachate, (ii) act as barrier to prevent direct contact with the contaminated soil, and (iii) prevent soil erosion and the distribution of the toxic metal to other areas [46]. Phytostabilization can occur through the process of sorption, precipitation, complexation, or metal valence reduction. This technique is useful for the cleanup of Pb, As, Cd, Cr, Cu, and Zn [147]. It can also be used to reestablish a plant community on sites that have been denuded due to the high levels of metal contamination. Once a community of tolerant species has been established, the potential for wind erosion (and thus spread of the pollutant) is reduced, and leaching of the soil contaminants is also reduced. Phytostabilization is advantageous because disposal of hazardous material/biomass is not required, and it is very effective when rapid immobilization is needed to preserve ground and surface waters [147, 148].

### *Phytofiltration*

Phytofiltration is the use of plant roots (rhizofiltration) or seedlings (blastofiltration), is similar in concept to phytoextraction, but is used to absorb or adsorb pollutants, mainly metals, from groundwater and aqueous-waste streams rather than the remediation of polluted soils [3, 123]. Rhizosphere is the soil area immediately surrounding the plant root surface, typically up to a few millimetres from the root surface. The contaminants are either adsorbed onto the root surface or are absorbed by the plant roots. Plants used for rhizofiltration are not planted directly in situ but are acclimated to the pollutant first. Plants are hydroponically grown in clean water rather than soil, until a large root system has developed. Once a large root system is in place, the water supply is substituted for a polluted water supply to acclimatize the plant. After the plants become acclimatized, they are planted in the polluted area where the roots uptake the polluted water and the contaminants along with it. As the roots become saturated, they are harvested and disposed of safely. Repeated treatments of the site can reduce pollution to suitable levels as was exemplified in Chernobyl where sunflowers were grown in radioactively contaminated pools [21].

## CONCLUSIONS

Background knowledge of the sources, chemistry, and potential risks of toxic heavy metals in contaminated soils is necessary for the selection of appropriate remedial options. Remediation of soil contaminated by heavy metals is necessary in order to reduce the associated risks, make the land resource available for agricultural production, enhance food security, and scale down land tenure problems. Immobilization, soil washing, and phytoremediation are frequently listed among the best available technologies for cleaning up heavy metal contaminated soils but have been mostly demonstrated in developed countries. These technologies are recommended for field applicability and commercialization in the developing countries also where agriculture, urbanization, and industrialization are leaving a legacy of environmental degradation.

# REFERENCES

1. S. Khan, Q. Cao, Y. M. Zheng, Y. Z. Huang, and Y. G. Zhu, "Health risks of heavy metals in contaminated soils and food crops irrigated with wastewater in Beijing, China," Environmental Pollution, vol. 152, no. 3, pp. 686–692, 2008.
2. M. K. Zhang, Z. Y. Liu, and H. Wang, "Use of single extraction methods to predict bioavailability of heavy metals in polluted soils to rice," Communications in Soil Science and Plant Analysis, vol. 41, no. 7, pp. 820–831, 2010.
3. GWRTAC, "Remediation of metals-contaminated soils and groundwater," Tech. Rep. TE-97-01,, GWRTAC, Pittsburgh, Pa, USA, 1997, GWRTAC-E Series.
4. T. A. Kirpichtchikova, A. Manceau, L. Spadini, F. Panfili, M. A. Marcus, and T. Jacquet, "Speciation and solubility of heavy metals in contaminated soil using X-ray microfluorescence, EXAFS spectroscopy, chemical extraction, and thermodynamic modeling," Geochimica et Cosmochimica Acta, vol. 70, no. 9, pp. 2163–2190, 2006.
5. D. C. Adriano, Trace Elements in Terrestrial Environments: Biogeochemistry, Bioavailability and Risks of Metals, Springer, New York, NY, USA, 2nd edition, 2003.
6. P. Maslin and R. M. Maier, "Rhamnolipid-enhanced mineralization of phenanthrene in organic-metal co-contaminated soils," Bioremediation Journal, vol. 4, no. 4, pp. 295–308, 2000.
7. M. J. McLaughlin, B. A. Zarcinas, D. P. Stevens, and N. Cook, "Soil testing for heavy metals,"Communications in Soil Science and Plant Analysis, vol. 31, no. 11–14, pp. 1661–1700, 2000.
8. M. J. McLaughlin, R. E. Hamon, R. G. McLaren, T. W. Speir, and S. L. Rogers, "Review: a bioavailability-based rationale for controlling metal and metalloid contamination of agricultural land in Australia and New Zealand," Australian Journal of Soil Research, vol. 38, no. 6, pp. 1037–1086, 2000.
9. W. Ling, Q. Shen, Y. Gao, X. Gu, and Z. Yang, "Use of bentonite to control the release of copper from contaminated soils," Australian Journal of Soil Research, vol. 45, no. 8, pp. 618–623, 2007.

10. A. Kabata-Pendias and H. Pendias, Trace Metals in Soils and Plants, CRC Press, Boca Raton, Fla, USA, 2nd edition, 2001.
11. Q. Zhao and J. J. Kaluarachchi, "Risk assessment at hazardous waste-contaminated sites with variability of population characteristics," Environment International, vol. 28, no. 1-2, pp. 41–53, 2002.
12. N. S. Bolan, B.G. Ko, C.W.N. Anderson, and I. Vogeler, "Solute interactions in soils in relation to bioavailability and remediation of the environment," in Proceedings of the 5th International Symposium of Interactions of Soil Minerals with Organic Components and Microorganisms, Pucón, Chile, November 2008.
13. G. M. Pierzynski, J. T. Sims, and G. F. Vance, Soils and Environmental Quality, CRC Press, London, UK, 2nd edition, 2000.
14. J. J. D'Amore, S. R. Al-Abed, K. G. Scheckel, and J. A. Ryan, "Methods for speciation of metals in soils: a review," Journal of Environmental Quality, vol. 34, no. 5, pp. 1707–1745, 2005.
15. B. J. Alloway, Heavy Metals in Soils, Blackie Academic and Professional, London, UK, 2nd edition, 1995.
16. E. Lombi and M. H. Gerzabek, "Determination of mobile heavy metal fraction in soil: results of a pot experiment with sewage sludge," Communications in Soil Science and Plant Analysis, vol. 29, no. 17-18, pp. 2545–2556, 1998.
17. G. Sposito and A. L. Page, "Cycling of metal ions in the soil environment," in Metal Ions in Biological Systems, H. Sigel, Ed., vol. 18 of Circulation of Metals in the Environment, pp. 287–332, Marcel Dekker, Inc., New York, NY, USA, 1984.
18. S. Kuo, P. E. Heilman, and A. S. Baker, "Distribution and forms of copper, zinc, cadmium, iron, and manganese in soils near a copper smelter," Soil Science, vol. 135, no. 2, pp. 101–109, 1983.
19. M. Kaasalainen and M. Yli-Halla, "Use of sequential extraction to assess metal partitioning in soils,"Environmental Pollution, vol. 126, no. 2, pp. 225–233, 2003.

20. N. T. Basta, J. A. Ryan, and R. L. Chaney, "Trace element chemistry in residual-treated soil: key concepts and metal bioavailability," Journal of Environmental Quality, vol. 34, no. 1, pp. 49–63, 2005.
21. A. Scragg, Environmental Biotechnology, Oxford University Press, Oxford, UK, 2nd edition, 2006.
22. M.M. Lasat, "Phytoextraction of metals from contaminated soil: a review of plant/soil/metal interaction and assessment of pertinent agronomic issues," Journal of Hazardous Substances Research, vol. 2, pp. 1–25, 2000.
23. L. H. P. Jones and S. C. Jarvis, "The fate of heavy metals," in The Chemistry of Soil Processes, D. J. Green and M. H. B. Hayes, Eds., p. 593, John Wiley & Sons, New York, NY, USA, 1981.
24. P. H. Raven, L. R. Berg, and G. B. Johnson, Environment, Saunders College Publishing, New York, NY, USA, 2nd edition, 1998.
25. M. E. Sumner, "Beneficial use of effluents, wastes, and biosolids," Communications in Soil Science and Plant Analysis, vol. 31, no. 11–14, pp. 1701–1715, 2000.
26. R. L. Chaney and D. P. Oliver, "Sources, potential adverse effects and remediation of agricultural soil contaminants," in Contaminants and the Soil Environments in the Australia-Pacific Region, R. Naidu, Ed., pp. 323–359, Kluwer Academic Publishers, Dordrecht, The Netherlands, 1996.
27. USEPA, "A plain english guide to the EPA part 503 biosolids rule," USEPA Rep. 832/R-93/003, USEPA, Washington, DC, USA, 1994.
28. K. Weggler, M. J. McLaughlin, and R. D. Graham, "Effect of Chloride in Soil Solution on the Plant Availability of Biosolid-Borne Cadmium," Journal of Environmental Quality, vol. 33, no. 2, pp. 496–504, 2004.
29. M. L. A. Silveira, L. R. F. Alleoni, and, and L. R. G. Guilherme, "Biosolids and heavy metals in soils,"Scientia Agricola, vol. 60, no. 4, pp. 64–111, 2003.
30. R. Canet, F. Pomares, F. Tarazona, and M. Estela, "Sequential fractionation and plant availability of heavy metals as affected by sewage sludge applications to soil," Communications in Soil Science and Plant Analysis, vol. 29, no. 5-6, pp. 697–716, 1998.

31. S. V. Mattigod and A. L. Page, "Assessment of metal pollution in soil," in Applied Environmental Geochemistry, pp. 355–394, Academic Press, London, UK, 1983.

32. R. G. McLaren, L. M. Clucas, and M. D. Taylor, "Leaching of macronutrients and metals from undisturbed soils treated with metal-spiked sewage sludge. 3. Distribution of residual metals," Australian Journal of Soil Research, vol. 43, no. 2, pp. 159–170, 2005.

33. C. Keller, S. P. McGrath, and S. J. Dunham, "Trace metal leaching through a soil-grassland system after sewage sludge application," Journal of Environmental Quality, vol. 31, no. 5, pp. 1550–1560, 2002.

34. R. G. McLaren, L. M. Clucas, M. D. Taylor, and T. Hendry, "Leaching of macronutrients and metals from undisturbed soils treated with metal-spiked sewage sludge. 2. Leaching of metals," Australian Journal of Soil Research, vol. 42, no. 4, pp. 459–471, 2004.

35. S. C. Reed, R. W. Crites, and E. J. Middlebrooks, Natural Systems for Waste Management and Treatment, McGraw-Hill, New York, NY, USA, 2nd edition, 1995.

36. J. Bjuhr, Trace Metals in Soils Irrigated with Waste Water in a Periurban Area Downstream Hanoi City, Vietnam, Seminar Paper, Institutionen för markvetenskap, Sveriges lantbruksuniversitet (SLU), Uppsala, Sweden, 2007.

37. P. S. DeVolder, S. L. Brown, D. Hesterberg, and K. Pandya, "Metal bioavailability and speciation in a wetland tailings repository amended with biosolids compost, wood ash, and sulfate," Journal of Environmental Quality, vol. 32, no. 3, pp. 851–864, 2003.

38. N. T. Basta and R. Gradwohl, "Remediation of heavy metal-contaminated soil using rock phosphate,"Better Crops, vol. 82, no. 4, pp. 29–31, 1998.

39. L. A. Smith, J. L. Means, A. Chen, et al., Remedial Options for Metals-Contaminated Sites, Lewis Publishers, Boca Raton, Fla, USA,, 1995.

40. USEPA, Report: recent Developments for In Situ Treatment of Metals contaminated Soils, U.S. Environmental Protection Agency, Office of Solid Waste and Emergency Response, 1996.

41. J. Shiowatana, R. G. McLaren, N. Chanmekha, and A. Samphao, "Fractionation of arsenic in soil by a continuous-flow sequential extraction method," Journal of Environmental Quality, vol. 30, no. 6, pp. 1940–1949, 2001.

42. J. Buekers, Fixation of cadmium, copper, nickel and zinc in soil: kinetics, mechanisms and its effect on metal bioavailability, Ph.D. thesis, Katholieke Universiteit Lueven, 2007, Dissertationes De Agricultura, Doctoraatsprooefschrift nr.

43. D. B. Levy, K. A. Barbarick, E. G. Siemer, and L. E. Sommers, "Distribution and partitioning of trace metals in contaminated soils near Leadville, Colorado," Journal of Environmental Quality, vol. 21, no. 2, pp. 185–195, 1992.

44. USDHHS, Toxicological profile for lead, United States Department of Health and Human Services, Atlanta, Ga, USA, 1999.

45. S.E. Manahan, Toxicological Chemistry and Biochemistry, CRC Press, Limited Liability Company (LLC), 3rd edition, 2003.

46. I. Raskin and B. D. Ensley, Phytoremediation of Toxic Metals:Using Plants to Clean Up the Environment, John Wiley & Sons, New York, NY, USA, 2000.

47. NSC, Lead Poisoning, National Safety Council, 2009,http://www.nsc.org/news_resources/Resources/Documents/Lead_Poisoning.pdf.

48. D. R. Baldwin and W. J. Marshall, "Heavy metal poisoning and its laboratory investigation," Annals of Clinical Biochemistry, vol. 36, no. 3, pp. 267–300, 1999.

49. C.J. Rosen, Lead in the home garden and urban soil environment, Communication and Educational Technology Services, University of Minnesota Extension, 2002.

50. P. Chrostowski, J. L. Durda, and K. G. Edelmann, "The use of natural processes for the control of chromium migration," Remediation, vol. 2, no. 3, pp. 341–351, 1991.

51. I. Bodek, W. J. Lyman, W. F. Reehl, and D. H. Rosenblatt, in Environmental Inorganic Chemistry: Properties, Processes and Estimation Methods, Pergamon Press, Elmsford, NY, USA, 1988.

52. B. E. Davies and L. H. P. Jones, "Micronutrients and toxic elements," in Russell's Soil Conditions and Plant Growth, A.

Wild, Ed., pp. 781–814, John Wiley & Sons; Interscience, New York, NY, USA, 11th edition, 1988.

53. K. M. Greany, An assessment of heavy metal contamination in the marine sediments of Las Perlas Archipelago, Gulf of Panama, M.S. thesis, School of Life Sciences Heriot-Watt University, Edinburgh, Scotland, 2005.
54. P. G. C. Campbell, "Cadmium-A priority pollutant," Environmental Chemistry, vol. 3, no. 6, pp. 387–388, 2006.
55. VCI, Copper history/Future, Van Commodities Inc., 2011,http://trademetalfutures.com/copperhistory.html.
56. C. E. Martínez and H. L. Motto, "Solubility of lead, zinc and copper added to mineral soils,"Environmental Pollution, vol. 107, no. 1, pp. 153–158, 2000.
57. J. Eriksson, A. Andersson, and R. Andersson, "The state of Swedish farmlands," Tech. Rep. 4778, Swedish Environmental Protection Agency, Stockholm, Sweden, 1997.
58. M. Pourbaix, Atlas of Electrochemical Equilibria, Pergamon Press, New York, NY, USA, 1974, Translated from French by J.A. Franklin.
59. A. P. Khodadoust, K. R. Reddy, and K. Maturi, "Removal of nickel and phenanthrene from kaolin soil using different extractants," Environmental Engineering Science, vol. 21, no. 6, pp. 691–704, 2004.
60. DPR-EGASPIN, Environmental Guidelines and Standards for the Petroleum Industry in Nigeria (EGASPIN), Department of Petroleum Resources, Lagos, Nigeria, 2002.
61. A. Tessier, P. G. C. Campbell, and M. Blsson, "Sequential extraction procedure for the speciation of particulate trace metals," Analytical Chemistry, vol. 51, no. 7, pp. 844–851, 1979.
62. A. M. Ure, PH. Quevauviller, H. Muntau, and B. Griepink, "Speciation of heavy metals in soils and sediments. An account of the improvement and harmonization of extraction techniques undertaken under the auspices of the BCR of Commission of the European Communities," International Journal of Environmental Analytical Chemistry, vol. 51, no. 1, pp. 35–151, 1993.
63. D. M. DiToro, J. D. Mahony, D. J. Hansen, K. J. Scott, A. R. Carlson, and G. T. Ankley, "Acid volatile sulfide predicts the acute toxicity

of cadmium and nickel in sediments," Environmental Science and Technology, vol. 26, no. 1, pp. 96–101, 1992.

64. R. G. Riley, J. M. Zachara, and F. J. Wobber, "Chemical contaminants on DOE lands and selection of contaminated mixtures for subsurface science research," US-DOE, Energy Resource Subsurface Science Program, Washington, DC, USA, 1992.

65. NJDEP, Soil Cleanup Criteria, New Jersey Department of Environmental Protection, Proposed Cleanup Standards for Contaminated Sites, NJAC 7:26D, 1996.

66. T. A. Martin and M. V. Ruby, "Review of in situ remediation technologies for lead, zinc and cadmium in soil," Remediation, vol. 14, no. 3, pp. 35–53, 2004.

67. S. K. Gupta, T. Herren, K. Wenger, R. Krebs, and T. Hari, "In situ gentle remediation measures for heavy metal-polluted soils," in Phytoremediation of Contaminated Soil and Water, N. Terry and G. Bañuelos, Eds., pp. 303–322, Lewis Publishers, Boca Raton, Fla, USA, 2000.

68. USEPA, "Treatment technologies for site cleanup: annual status report (12th Edition)," Tech. Rep. EPA-542-R-07-012, Solid Waste and Emergency Response (5203P), Washington, DC, USA, 2007.

69. USEPA, "Recent developments for in situ treatment of metal contaminated soils," Tech. Rep. EPA-542-R-97-004, USEPA, Washington, DC, USA, 1997.

70. Y. Hashimoto, H. Matsufuru, M. Takaoka, H. Tanida, and T. Sato, "Impacts of chemical amendment and plant growth on lead speciation and enzyme activities in a shooting range soil: an X-ray absorption fine structure investigation," Journal of Environmental Quality, vol. 38, no. 4, pp. 1420–1428, 2009.

71. N. Finžgar, B. Kos, and D. Leštan, "Bioavailability and mobility of Pb after soil treatment with different remediation methods," Plant, Soil and Environment, vol. 52, no. 1, pp. 25–34, 2006.

72. J. Boisson, M. Mench, J. Vangronsveld, A. Ruttens, P. Kopponen, and T. De Koe, "Immobilization of trace metals and arsenic by different soil additives: evaluation by means of chemical extractions,"Communications in Soil Science and Plant Analysis, vol. 30, no. 3-4, pp. 365–387, 1999.

73. E. Lombi, F. J. Zhao, G. Zhang et al., "In situ fixation of metals in soils using bauxite residue: chemical assessment," Environmental Pollution, vol. 118, no. 3, pp. 435–443, 2002.

74. C. O. Anoduadi, L. B. Okenwa, F. E. Okieimen, A. T. Tyowua, and E.G. Uwumarongie-Ilori, "Metal immobilization in CCA contaminated soil using laterite and termite mound soil. Evaluation by chemical fractionation," Nigerian Journal of Applied Science, vol. 27, pp. 77–87, 2009.

75. L. Q. Wang, L. Luo, Y. B Ma, D. P. Wei, and L. Hua, "In situ immobilization remediation of heavy metals-contaminated soils: a review," Chinese Journal of Applied Ecology, vol. 20, no. 5, pp. 1214–1222, 2009.

76. F. R. Evanko and D. A. Dzombak, "Remediation of metals contaminated soils and groundwater," Tech. Rep. TE-97-01, Groundwater Remediation Technologies Analysis Centre, Pittsburg, Pa, USA, 1997.

77. E. M. Fawzy, "Soil remediation using in situ immobilisation techniques," Chemistry and Ecology, vol. 24, no. 2, pp. 147–156, 2008.

78. M. Farrell, W. T. Perkins, P. J. Hobbs, G. W. Griffith, and D. L. Jones, "Migration of heavy metals in soil as influenced by compost amendments," Environmental Pollution, vol. 158, no. 1, pp. 55–64, 2010.

79. P. Bishop, D. Gress, and J. Olafsson, "Cement stabilization of heavy metals:Leaching rate assessment," inIndustrial Wastes-Proceedings of the 14th Mid-Atlantic Industrial Waste Conference, Technomics, Lancaster, Pa, USA, 1982.

80. W. Shively, P. Bishop, D. Gress, and T. Brown, "Leaching tests of heavy metals stabilized with Portland cement," Journal of the Water Pollution Control Federation, vol. 58, no. 3, pp. 234–241, 1986.

81. USEPA, "Interference mechanisms in waste stabilization/solidification processes," Tech. Rep. EPA/540/A5-89/004, United States Environmental Protection Agency, Office of Research and Development, Cincinnati, Ohio, USA, 1990.

82. G. Guo, Q. Zhou, and L. Q. Ma, "Availability and assessment of fixing additives for the in situ remediation of heavy metal

contaminated soils: a review," Environmental Monitoring and Assessment, vol. 116, no. 1–3, pp. 513–528, 2006.

83. J. R. Conner, Chemical Fixation and Solidification of Hazardous Wastes, Van Nostrand Reinhold, New York, NY, USA, 1990.

84. USEPA, "Stabilization/solidification of CERCLA and RCRA wastes," Tech. Rep. EPA/625/6-89/022, United States Environmental Protection Agency, Center for Environmental Research Information, Cincinnati, Ohio, USA, 1989.

85. USEPA, "International waste technologies/geo-con in situ stabilization/solidification," Tech. Rep. EPA/540/A5-89/004, United States Environmental Protection Agency, Office of Research and Development, Cincinnati, Ohio, USA, 1990.

86. B. H. Jasperse and C. R. Ryan, "Stabilization and fixation using soil mixing," in Proceedings of the ASCE Specialty Conference on Grouting, Soil Improvement, and Geosynthetics, ASCE Publications, Reston, Va, USA, 1992.

87. C. R. Ryan and A. D. Walker, "Soil mixing for soil improvement," in Proceedings of the 23rd Conference on In situ Soil Modification, Geo-Con, Inc., Louisville, Ky, USA, 1992.

88. USEPA, "Vitrification technologies for treatment of Hazardous and radioactive waste handbook," Tech. Rep. EPA/625/R-92/002, United States Environmental Protection Agency, Office of Research and Development, Washington, DC, USA, 1992.

89. J. L. Buelt and L. E. Thompson, The In situ Vitrification Integrated Program: Focusing on an Innovative Solution on Environmental Restoration Needs, Battelle Pacific Northwest Laboratory, Richland, Wash, USA, 1992.

90. A. Jang, Y. S. Choi, and I. S. Kim, "Batch and column tests for the development of an immobilization technology for toxic heavy metals in contaminated soils of closed mines," Water Science and Technology, vol. 37, no. 8, pp. 81–88, 1998.

91. G. Dermont, M. Bergeron, G. Mercier, and M. Richer-Laflèche, "Soil washing for metal removal: a review of physical/chemical technologies and field applications," Journal of Hazardous Materials, vol. 152, no. 1, pp. 1–31, 2008.

92. P. Wood, "Remediation methods for contaminated sites," in Contaminated Land and Its Reclamation, R. Hester and R.

Harrison, Eds., Royal Society of Chemistry, Cambridge, UK, 1997.

93. GOC, "Site Remediation Technologies: A Reference Manual," 2003, Contaminated Sites Working Group, Government of Canada, Ontario, Canada.
94. R. W. Peters, "Chelant extraction of heavy metals from contaminated soils," Journal of Hazardous Materials, vol. 66, no. 1-2, pp. 151–210, 1999.
95. CLAIRE, "Understanding soil washing, contaminated land: applications in real environments," Tech. Rep. TB13, 2007.
96. M. Pearl and P. Wood, "Review of pilot and full scale soil washing plants," Warren Spring Laboratory Report LR 1018, Department of the Environment, AEA Technology National Environmental Technology Centre, 1994, B551 Harwell, Oxfordshire, OX11 0RA.
97. A. Gosselin, M. Blackburn, and M. Bergeron, Assessment Protocol of the applicability of ore-processing technology to Treat Contaminated Soils, Sediments and Sludges, prepared for Eco-Technology innovation Section, Eco-Technology Innovation Section, Technology Development and Demonstration Program, Environment Canada, Canada, 1999.
98. A. P. Davis and I. Singh, "Washing of zinc(II) from contaminated soil column," Journal of Environmental Engineering, vol. 121, no. 2, pp. 174–185, 1995.
99. D. Gombert, "Soil washing and radioactive contamination," Environmental Progress, vol. 13, no. 2, pp. 138–142, 1994.
100. R. S. Tejowulan and W. H. Hendershot, "Removal of trace metals from contaminated soils using EDTA incorporating resin trapping techniques," Environmental Pollution, vol. 103, no. 1, pp. 135–142, 1998.
101. USEPA, "Engineering bulletin: soil washing treatment," Tech. Rep. EPA/540/2-90/017, Office of Emergency and Remedial Response, United States Environmental Protection Agency, Washington, DC, USA, 1990.
102. A. L. Wood, D. C. Bouchard, M. L. Brusseau, and P. S. C. Rao, "Cosolvent effects on sorption and mobility of organic contaminants in soils," Chemosphere, vol. 21, no. 4-5, pp. 575–587, 1990.

103. W. Chu and K. H. Chan, "The mechanism of the surfactant-aided soil washing system for hydrophobic and partial hydrophobic organics," Science of the Total Environment, vol. 307, no. 1–3, pp. 83–92, 2003.
104. Y. Gao, J. He, W. Ling, H. Hu, and F. Liu, "Effects of organic acids on copper and cadmium desorption from contaminated soils," Environment International, vol. 29, no. 5, pp. 613–618, 2003.
105. K. Maturi and K. R. Reddy, "Extractants for the removal of mixed contaminants from soils," Soil and Sediment Contamination, vol. 17, no. 6, pp. 586–608, 2008.
106. H. Zhang, Z. Dang, L. C. Zheng, and X. Y. Yi, "Remediation of soil co-contaminated with pyrene and cadmium by growing maize (Zea mays L.)," International Journal of Environmental Science and Technology, vol. 6, no. 2, pp. 249–258, 2009.
107. J. Yu and D. Klarup, "Extraction kinetics of copper, zinc, iron, and manganese from contaminated sediment using disodium ethylenediaminetetraacetate," Water, Air, and Soil Pollution, vol. 75, no. 3-4, pp. 205–225, 1994.
108. R. Naidu and R. D. Harter, "Effect of different organic ligands on cadmium sorption by and extractability from soils," Soil Science Society of America Journal, vol. 62, no. 3, pp. 644–650, 1998.
109. J. Labanowski, F. Monna, A. Bermond et al., "Kinetic extractions to assess mobilization of Zn, Pb, Cu, and Cd in a metal-contaminated soil: EDTA vs. citrate," Environmental Pollution, vol. 152, no. 3, pp. 693–701, 2008.
110. X. Ke, P. J. Li, Q. X. Zhou, Y. Zhang, and T. H. Sun, "Removal of heavy metals from a contaminated soil using tartaric acid," Journal of Environmental Sciences, vol. 18, no. 4, pp. 727–733, 2006.
111. B. Sun, F. J. Zhao, E. Lombi, and S. P. McGrath, "Leaching of heavy metals from contaminated soils using EDTA," Environmental Pollution, vol. 113, no. 2, pp. 111–120, 2001.
112. R. A. Wuana, F. E. Okieimen, and J. A. Imborvungu, "Removal of heavy metals from a contaminated soil using organic chelating acids," International Journal of Environmental Science and Technology, vol. 7, no. 3, pp. 485–496, 2010.

113. H. Farrah and W. F. Pickering, "Extraction of heavy metal ions sorbed on clays," Water, Air, and Soil Pollution, vol. 9, no. 4, pp. 491–498, 1978.

114. B. J. W. Tuin and M. Tels, "Removing heavy metals from contaminated clay soils by extraction with hydrochloric acid, edta or hypochlorite solutions," Environmental Technology, vol. 11, no. 11, pp. 1039–1052, 1990.

115. K. R. Reddy and S. Chinthamreddy, "Comparison of extractants for removing heavy metals from contaminated clayey soils," Soil and Sediment Contamination, vol. 9, no. 5, pp. 449–462, 2000.

116. A. P. Khodadoust, K. R. Reddy, and K. Maturi, "Effect of different extraction agents on metal and organic contaminant removal from a field soil," Journal of Hazardous Materials, vol. 117, no. 1, pp. 15–24, 2005.

117. T. C. Chen and A. Hong, "Chelating extraction of lead and copper from an authentic contaminated soil using N-(2-acetamido) iminodiacetic acid and S-carboxymethyl-L-cysteine," Journal of Hazardous Materials, vol. 41, no. 2-3, pp. 147–160, 1995.

118. R. A. Wuana, F. E. Okicimen, and R. E. Ikyereve, "Removal of lead and copper from contaminated kaolin and bulk clay soils using acids and chelating agents," Journal of Chemical Society of Nigeria, vol. 33, no. 1, pp. 213–219, 2008.

119. S. D. Cunningham and D. W. Ow, "Promises and prospects of phytoremediation," Plant Physiology, vol. 110, no. 3, pp. 715–719, 1996.

120. H. S. Helmisaari, M. Salemaa, J. Derome, O. Kiikkilä, C. Uhlig, and T. M. Nieminen, "Remediation of heavy metal-contaminated forest soil using recycled organic matter and native woody plants," Journal of Environmental Quality, vol. 36, no. 4, pp. 1145–1153, 2007.

121. R. L. Chaney, M. Malik, Y. M. Li et al., "Phytoremediation of soil metals," Current Opinion in Biotechnology, vol. 8, no. 3, pp. 279–284, 1997.

122. R. J. Henry, An Overview of the Phytoremediation of Lead and Mercury, United States Environmental Protection Agency Office of Solid Waste and Emergency Response Technology Innovation office, Washington, DC, USA, 2000.

123. C. Garbisu and I. Alkorta, "Phytoextraction: a cost-effective plant-based technology for the removal of metals from the environment," Bioresource Technology, vol. 77, no. 3, pp. 229–236, 2001.

124. C. D. Jadia and M. H. Fulekar, "Phytotoxicity and remediation of heavy metals by fibrous root grass (sorghum)," Journal of Applied Biosciences, vol. 10, no. 1, pp. 491–499, 2008.

125. M. Vysloužilová, P. Tlustoš, J. Száková, and D. Pavlíková, "As, Cd, Pb and Zn uptake by Salix spp. clones grown in soils enriched by high loads of these elements," Plant, Soil and Environment, vol. 49, no. 5, pp. 191–196, 2003.

126. E. Lombi, F. J. Zhao, S. J. Dunham, and S. P. McGrath, "Phytoremediation of heavy metal-contaminated soils: natural hyperaccumulation versus chemically enhanced phytoextraction," Journal of Environmental Quality, vol. 30, no. 6, pp. 1919–1926, 2001.

127. M. Ghosh and S. P. Singh, "A review on phytoremediation of heavy metals and utilization of its byproducts," Applied Ecology and Environmental Research, vol. 3, no. 1, pp. 1–18, 2005.

128. A. J. M. Baker and R. R. Brooks, "Terrestrial higher plants which hyperaccumulate metallic elements: a review of their distribution, ecology and phytochemistry," Biorecovery, vol. 1, pp. 81–126, 1989.

129. M. M. Lasat, "Phytoextraction of toxic metals: a review of biological mechanisms," Journal of Environmental Quality, vol. 31, no. 1, pp. 109–120, 2002.

130. D. E. Salt, R. D. Smith, and I. Raskin, "Phytoremediation," Annual Reviews in Plant Physiology & Plant Molecular Biology, vol. 49, pp. 643–668, 1998.

131. S. Dushenkov, "Trends in phytoremediation of radionuclides," Plant and Soil, vol. 249, no. 1, pp. 167–175, 2003.

132. U. Schmidt, "Enhancing phytoextraction: the effect of chemical soil manipulation on mobility, plant accumulation and leaching of heavy metals," Journal of Environmental Quality, vol. 32, no. 6, pp. 1939–1954, 2003.

133. B. Nowack, R. Schulin, and B. H. Robinson, "Critical assessment of chelant-enhanced metal phytoextraction," Environmental

Science and Technology, vol. 40, no. 17, pp. 5225–5232, 2006.

134. M. W. H. Evangelou, M. Ebel, and A. Schaeffer, "Chelate assisted phytoextraction of heavy metals from soil. Effect, mechanism, toxicity, and fate of chelating agents," Chemosphere, vol. 68, no. 6, pp. 989–1003, 2007.

135. J. W. Huang, J. Chen, W. R. Berti, and S. D. Cunningham, "Phytoremediadon of lead-contaminated soils: role of synthetic chelates in lead phytoextraction," Environmental Science and Technology, vol. 31, no. 3, pp. 800–805, 1997.

136. Saifullah, E. Meers, M. Qadir, et al., "EDTA-assisted Pb phytoextraction," Chemosphere, vol. 74, no. 10, pp. 1279–1291, 2009.

137. Y. Xu, N. Yamaji, R. Shen, and J. F. Ma, "Sorghum roots are inefficient in uptake of EDTA-chelated lead," Annals of Botany, vol. 99, no. 5, pp. 869–875, 2007.

138. A. D. Vassil, Y. Kapulnik, I. Raskin, and D. E. Sait, "The role of EDTA in lead transport and accumulation by Indian mustard," Plant Physiology, vol. 117, no. 2, pp. 447–453, 1998.

139. B. Kos and D. Leštan, "Chelator induced phytoextraction and in situ soil washing of Cu," Environmental Pollution, vol. 132, no. 2, pp. 333–339, 2004.

140. S. Tandy, K. Bossart, R. Mueller et al., "Extraction of heavy metals from soils using biodegradable chelating agents," Environmental Science and Technology, vol. 38, no. 3, pp. 937–944, 2004.

141. R. R. Brooks, M. F. Chambers, L. J. Nicks, and B. H. Robinson, "Phytomining," Trends in Plant Science, vol. 3, no. 9, pp. 359–362, 1998.

142. P. Zhuang, Z. H. Ye, C. Y. Lan, Z. W. Xie, and W. S. Shu, "Chemically assisted phytoextraction of heavy metal contaminated soils using three plant species," Plant and Soil, vol. 276, no. 1-2, pp. 153–162, 2005.

143. X. Zhang, H. Xia, Z. Li, P. Zhuang, and B. Gao, "Potential of four forage grasses in remediation of Cd and Zn contaminated soils," Bioresource Technology, vol. 101, no. 6, pp. 2063–2066, 2010.

144. L. A. Newman, S. E. Strand, N. Choe et al., "Uptake and biotransformation of trichloroethylene by hybrid poplars,"

Environmental Science and Technology, vol. 31, no. 4, pp. 1062–1067, 1997.

145. P. V. R. Iyer, T. R. Rao, and P. D. Grover, Biomass Thermochemical Characterization Characterization, Indian Institute of Technology, Delhi, India, 3rd edition, 2002.

146. M.D. Hetland, J. R. Gallagher, D. J. Daly, D. J. Hassett, and L. V. Heebink, "Processing of plants used to phytoremediate lead-contaminated sites," A. Leeson, E. A. Forte, M. K. Banks, and V. S. Magar, Eds., pp. 129–136, Batelle Press.

147. C. D. Jadia and M. H. Fulekar, "Phytoremediation of heavy metals: recent techniques," African Journal of Biotechnology, vol. 8, no. 6, pp. 921–928, 2009.

148. USEPA, "Introduction to phytoremediation," Tech. Rep. EPA 600/R-99/107, United States Environmental Protection Agency, Office of Research and Development, Cincinnati, Ohio, USA, 2000.

Chapter 5

# Dryland Winter Wheat Yield, Grain Protein, and Soil Nitrogen Responses to Fertilizer and Biosolids Applications

Richard T. Koenig[1], Craig G. Cogger[2], and Andy I. Bary[2]

[1]Department of Crop and Soil Sciences, Washington State University, Pullman, WA 99164-6420, USA

[2]Department of Crop and Soil Sciences and Puyallup Research and Extension Center, Washington State University, 2606 West Pioneer Way, Puyallup, WA 98371, USA

## ABSTRACT

Applications of biosolids were compared to inorganic nitrogen (N) fertilizer for two years at three locations in eastern Washington State, USA, with diverse rainfall and soft white, hard red, and hard white winter wheat (Triticum aestivum L.) cultivars. High rates of inorganic N tended to reduce yields, while grain protein responses to N rate were positive and linear for all wheat market classes. Biosolids produced 0 to 1400 kg ha$^{-1}$ (0 to 47%) higher grain yields than inorganic N. Wheat may have responded positively to nutrients other than N in the biosolids or to a metered N supply that limited vegetative growth and the potential for moisture stress-induced reductions in grain yield in these dryland production systems. Grain protein content with biosolids was either equal to or below grain protein with inorganic N, likely due to dilution of grain N from the higher yields achieved with biosolids. Results indicate the potential to improve dryland winter wheat yields with biosolids compared to inorganic N alone, but perhaps not to increase grain protein concentration of hard wheat when biosolids are applied immediately before planting.

# INTRODUCTION

Biosolids are an effective and relatively safe source of nitrogen (N) for dryland wheat production [1–3]. Applied at agronomic rates, biosolids can supply sufficient N to maximize yield, as well as a host of other nutrients that can benefit crops in a rotational sequence [4, 5]. Determining appropriate agronomic application rates is paramount in balancing nutrient (mainly N) needs of wheat without increasing the risk of nitrate ($NO_3^-$) leaching. Considerable research has been devoted to this subject [3, 5].

In the inland Pacific Northwest (PNW) USA, soft white winter wheat is the predominant crop grown on over 2.75 million ha of mainly dryland (rainfed) cropland [6]. The majority of this wheat is exported and used to make unleavened products such as flat breads, noodles, and cakes [7]. Low-grain protein concentration (<10%) is desirable when producing unleavened products. High-grain protein concentration in soft white winter wheat has been a problem in the PNW due, in part, to high soil N levels [7]. Previous biosolids research in this area has shown that agronomic applications at or above rates required to maximize yield may produce undesirably high grain protein concentrations in soft white winter wheat [3, 5]. While high grain protein concentration is detrimental for soft wheat end uses, high protein is desirable in hard red and white winter wheats, with optimum targets of approximately

11.5 and 12.5%, respectively. Biosolids may be more appropriately suited to hard wheat production in dryland areas of the PNW, as is the case in the state of Colorado, USA [8].

The objective of this study was to determine if biosolids applied at agronomic rates are a more suitable source of N for producing dryland hard red and white wheat grain with a higher and more desirable protein concentration than soft white wheat grain.

# MATERIALS AND METHODS

Field studies were conducted at three locations in fall 2006 and again in separate but nearby (<200 m distant) locations in fall 2007. The locations were selected to represent three common but contrasting rainfed wheat production systems in eastern Washington State characterized by a Mediterranean climate and precipitation gradient of <300 to >600 mm year$^{-1}$. The Lind site (46° 58.3′ N, 118° 36.9′ W) typically receives 200 to 250 mm precipitation year$^{-1}$ and is in a two-year, winter wheat-tillage fallow rotation. The Davenport site (47° 39.2′ N, 118° 9′ W) typically receives 250 to 350 mm precipitation year$^{-1}$ and is in a three-year, winter wheat-spring wheat-chemical fallow (no-tillage) rotation. The Pullman site (46° 43.9′ N, 117° 10.8′ W) typically receives 500 to 600 mm precipitation year$^{-1}$ and is in a three-year, winter wheat-spring wheat-spring legume no-till rotation. Actual precipitation totals received at each site during this study are presented in Table 1. Each site was farmed uniformly for >2 years prior to these studies.

**Table 1:** Crop year precipitation and preplant soil test information for each site-year. Plant-available soil moisture is calculated by the commercial testing lab from gravimetric moisture content, an assumed bulk density of 1.2 g $cm^{-3}$ for silt loam soil textures, and an assumed permanent wilting point moisture content of 11% by volume

| Location† | Year | Sept 1 to Aug 31 crop year precipitation | Plant-available moisture | Residual N | Organic matter | pH | $NaHCO_3$-extractable P‡ | $NaHCO_3$-extractable K | Water-extractable |
|---|---|---|---|---|---|---|---|---|---|
| | | (mm) | (cm 1.5-$m^{-1}$) | (kg $ha^{-1}$) | (g $kg^{-1}$) | | (mg $kg^{-1}$) | (mg $kg^{-1}$) | (kg S $ha^{-1}$) |
| Lind | 2006 | 304 | 5.5 (0.3) | 141 (22)¶ | 8 (1.0) | 7.5 (0.1) | 6 (0.6) | 524 (31) | 16 (3.2) |
| | 2007 | 229 | 2.5 (0.4) | 189 (17) | 11 (0.7) | 6.8 (0.1) | 17 (3.2) | 469 (10) | 12 (1.5) |
| Davenport | 2006 | 336 | 8.5 (0.7) | 156 (41) | 31 (3.2) | 6.7 (0.5) | 12 (0.6) | 437 (49) | 9 (0.6) |
| | 2007 | 196 | 8.5 | 113 | 24 | 5.6 | 20 | 416 | 7 |
| Pullman | 2006 | 452 | 3.9 | 70 | 23 | 5.5 | 28 | 281 | 11 |
| | 2007 | 510 | 6.4 | 76 | 30 | 5.8 | 26 | 266 | 8 |

†Soil series and taxonomic names. Lind: Shano silt loam—coarse-silty, mixed, mesic, superactive Xeric Haplocambid (both years). Davenport: Mondovi silt loam—coarse-silty, mixed, mesic, superactive Cumulic Haploxeroll (2006-07).

Hanning silt loam—fine-silty, mixed, mesic, superactive Pachic Argixeroll (2007-08).

Pullman: Palouse silt loam—fine-silty, mixed, superactive Pachic Ultic Haploxeroll (both years).

‡Critical values for sufficient soil test P (bicarbonate method) are >16 mg $kg^{-1}$ [9].

§Critical values for sufficient soil test S are >22 to 34 kg $ha^{-1}$ [9].

[¶]Mean (standard deviation) of three replicate samples. If no standard deviation is indicated then only one, 5-point composite was collected from the study area.

At each location, preplant soil samples were collected in 0.3-m increments to a depth of 1.5 m to quantify plant-available soil moisture and residual N, as well as other soil properties, prior to establishing each study (Table 1). Inorganic N fertilizer was applied at rates intended to supply 0 to 150 or 200% of the standard agronomic recommendation [9] based on forecast yield potential and the initial soil test N at each location. Actual rates of application ranged from 0 to 112 kg N $ha^{-1}$ except at Pullman 2007-08, where rates ranged from 0 to 180 kg N $ha^{-1}$. The N source used was dry urea (46% N).

Biosolids sources included Class A dewatered cake (22% solids) from Tacoma, Wash, produced by thermophilic-mesophilic digestion, and Soundgro, Class A heat-dried (93% solids) biosolids from Pierce County, Wash. Biosolids rates were calculated to supply approximately 1× and 2× of the agronomic rate of N for these scenarios. Data points for biosolids treatments are graphed on inorganic N response functions (Figures 1 and 2) at the mid and high N rates for each site-year. Biosolids rates and equivalent plant-available N supply are presented in Table 2. Based on previous research on N availability from dewatered and heat-dried biosolids [10] we estimated plant-available N of both biosolids sources as 25% of total N under the conditions of this study and used that estimate for our field application rates (Table 2). In a more recent study of heat-dried biosolids conducted in western Washington State in 2007–2009, Cogger et al. [11] reported much greater N availability from Soundgro, and we adjusted our estimate of available N applied from Soundgro to 50% of total N. The ramifications of this change are discussed later. Biosolids were surface applied at or up to two weeks prior to planting at each location. Biosolids were not incorporated. Inorganic N was applied two weeks prior to planting at the Lind location and immediately prior to planting at Davenport and Pullman by banding the N 20 to 25 cm below the surface. At Davenport and Pullman, an ammonium phosphate-sulfate starter fertilizer was applied with the seed at rates of 9 kg N, 5 kg P, and 13 kg S $ha^{-1}$.

**Table 2:** Biosolids total N, ammonium $(NH_4)$–N, solids content and material, and N application rates for each site-year. Soundgro and Tacoma are two biosolids products produced in the Seattle, Washington metropolitan area

| Year | Site | Biosolids origin | Total N | $NH_4$–N | Solids | Rate code | Dry application rate | Total N | Estimated available N† | Plant-available N |
|---|---|---|---|---|---|---|---|---|---|---|
| | | | % | % | % | | Mg ha$^{-1}$ | kg ha$^{-1}$ | % of total N | Kg ha$^{-1}$ |
| 2006 | Lind | Soundgro | 5.80 | 0.53 | 93.3 | 1× | 4.4 | 257 | 50 | 129 |
| | | | 5.80 | 0.53 | 93.3 | 2× | 8.8 | 508 | 50 | 254 |
| | Davenport | Soundgro | 5.85 | 0.56 | 91.6 | 1× | 5.4 | 314 | 50 | 157 |
| | | | 5.85 | 0.56 | 91.6 | 2× | 10.9 | 635 | 50 | 317 |
| | Pullman | Tacoma | 4.60 | 0.78 | 21.5 | 1× | 6.4 | 295 | 25 | 74 |
| | | | 4.60 | 0.78 | 21.5 | 2× | 12.9 | 591 | 25 | 148 |
| 2007 | Lind | Tacoma | 4.90 | 0.40 | 22.5 | 1× | 5.0 | 247 | 25 | 62 |
| | | | 4.90 | 0.40 | 22.5 | 2× | 10.1 | 494 | 25 | 123 |
| | Wilke | Soundgro | 6.10 | 0.29 | 93.6 | 1× | 5.5 | 334 | 50 | 167 |
| | | | 6.10 | 0.29 | 93.6 | 2× | 11.1 | 676 | 50 | 338 |
| | Pullman | Soundgro | 6.10 | 0.29 | 93.6 | 1× | 6.5 | 398 | 50 | 199 |
| | | | 6.10 | 0.29 | 93.6 | 2× | 13.0 | 796 | 50 | 398 |

Available N estimate is 25% total N for Tacoma product and 50% total N for Soundgro based on Sullivan et al. [3] and Cogger et al. [11].

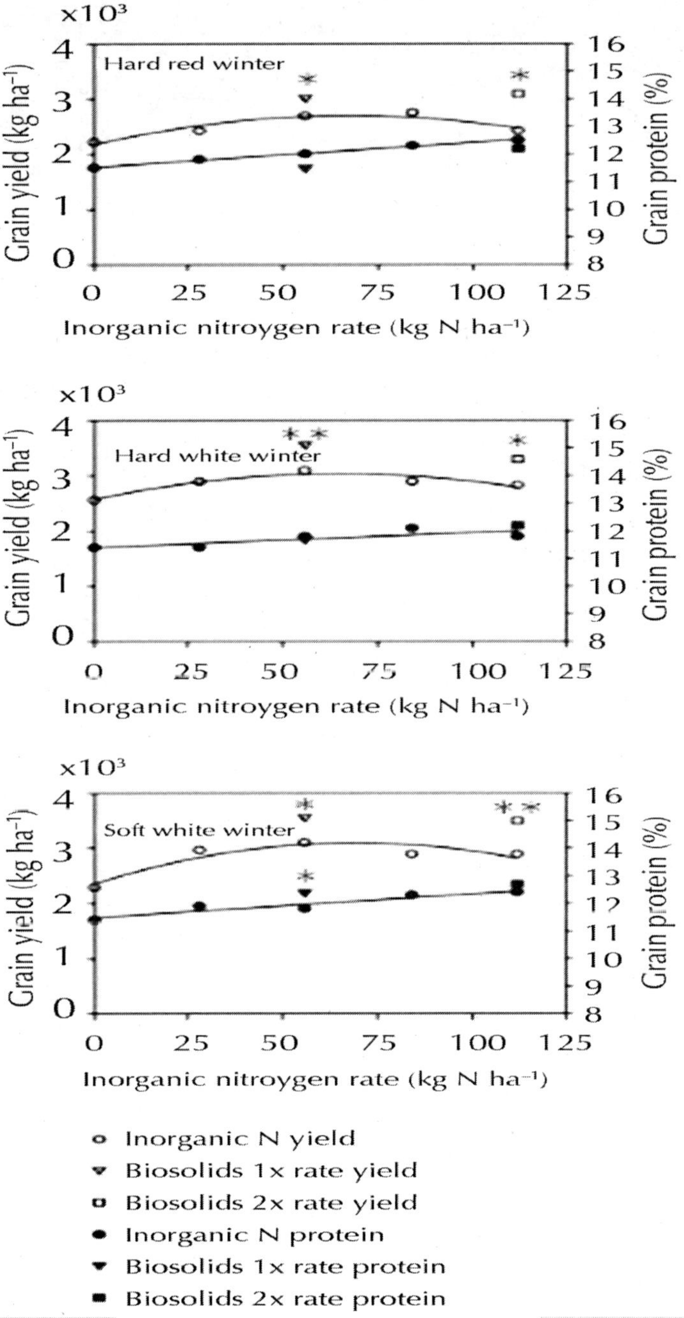
x10³
Hard red winter
Grain yield (kg ha⁻¹)
Grain protein (%)
Inorganic nitroygen rate (kg N ha⁻¹)
x10³
Hard white winter
Grain yield (kg ha⁻¹)
Grain protein (%)
Inorganic nitroygen rate (kg N ha⁻¹)
x10³
Soft white winter
Grain yield (kg ha⁻¹)
Grain protein (%)
Inorganic nitroygen rate (kg N ha⁻¹)
Inorganic N yield
Biosolids 1x rate yield
Biosolids 2x rate yield
Inorganic N protein
Biosolids 1x rate protein
Biosolids 2x rate protein

(a)

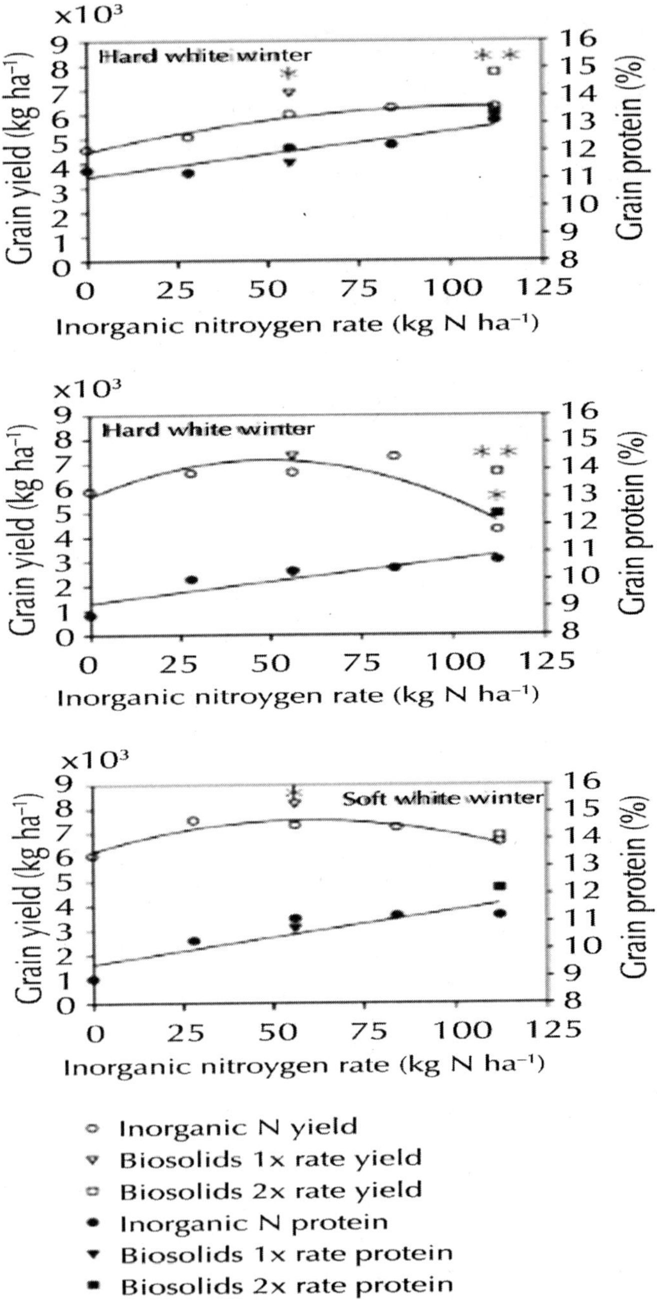
×10³
Hard white winter
Grain yield (kg ha⁻¹)
Grain protein (%)
Inorganic nitroygen rate (kg N ha⁻¹)
×10³
Hard white winter
Grain yield (kg ha⁻¹)
Grain protein (%)
Inorganic nitroygen rate (kg N ha⁻¹)
×10³
Soft white winter
Grain yield (kg ha⁻¹)
Grain protein (%)
Inorganic nitroygen rate (kg N ha⁻¹)
Inorganic N yield
Biosolids 1x rate yield
Biosolids 2x rate yield
Inorganic N protein
Biosolids 1x rate protein
Biosolids 2x rate protein

(b)

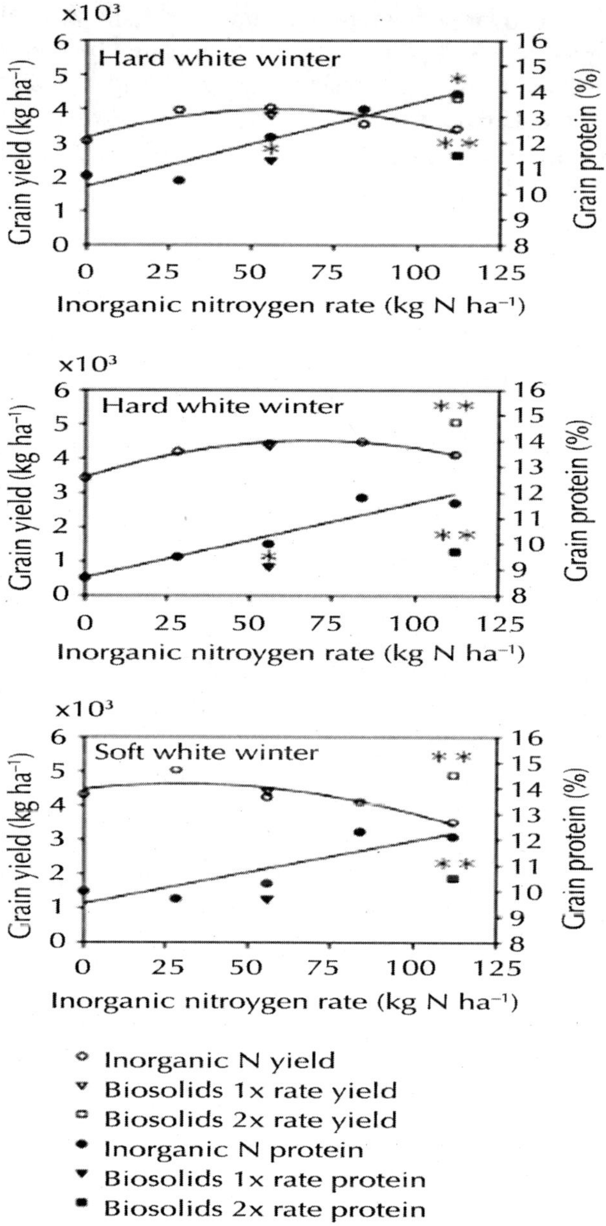


(c)

**Figure 1:** The effect of inorganic N fertilizer and biosolids on yield and grain protein content of three wheat market classes at Lind (a), Davenport (b), and

Pullman (c), Wash in 2007. The horizontal axis (N rate) is the same for each graph panel. Due to large differences in grain yield, the vertical axis changes for each market class site (column). Where indicated by the regression line, quadratic or linear responses are significant at. * And ** indicate significant differences at the 0.10 and 0.05 level, respectively, between inorganic N and biosolids treatments at that inorganic nitrogen and biosolids rate.

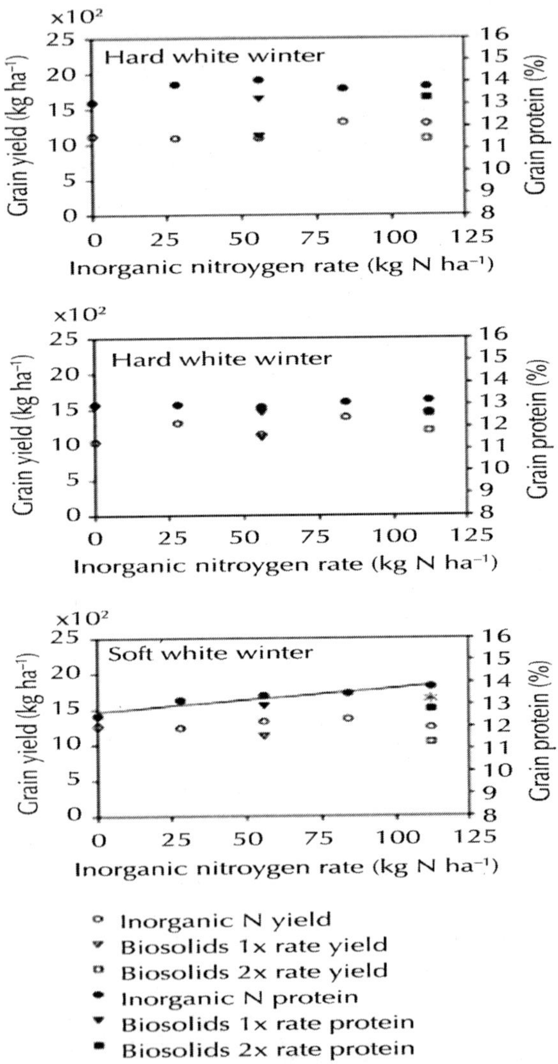


(d

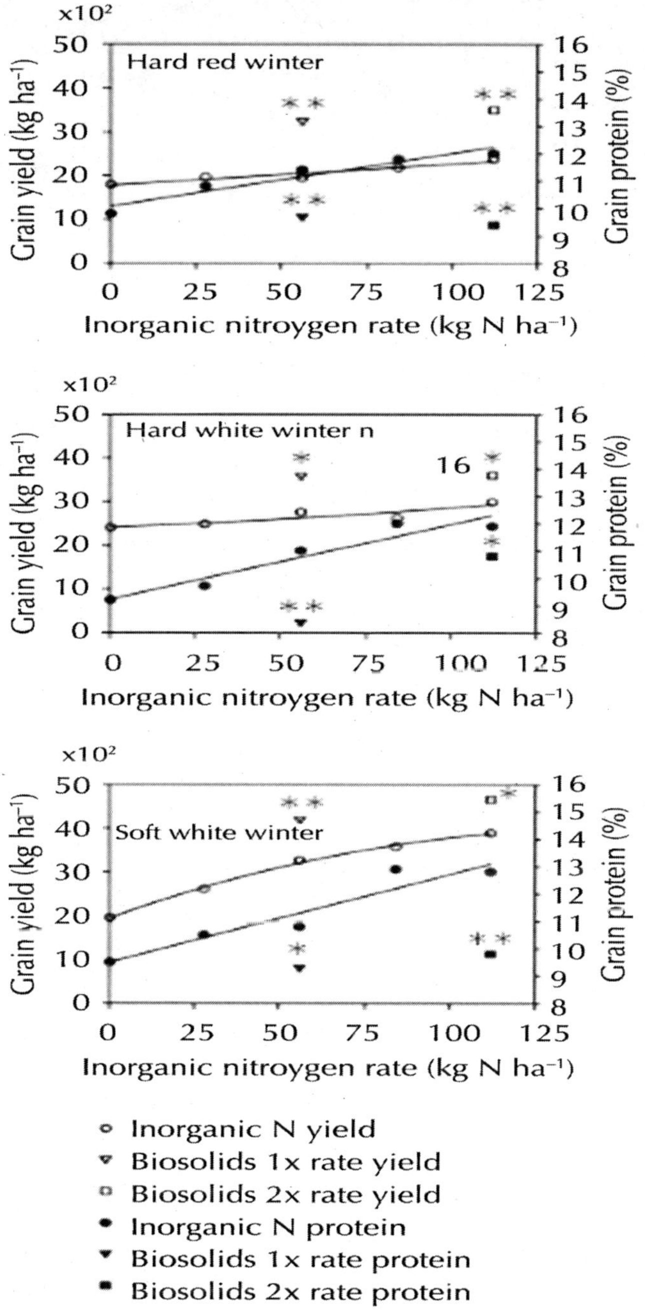
x10²
Grain yield (kg ha⁻¹)
Hard red winter
Grain protein (%)
Inorganic nitroygen rate (kg N ha⁻¹)
x10²
Grain yield (kg ha⁻¹)
Hard white winter n
Grain protein (%)
Inorganic nitroygen rate (kg N ha⁻¹)
x10²
Grain yield (kg ha⁻¹)
Soft white winter
Grain protein (%)
Inorganic nitroygen rate (kg N ha⁻¹)
Inorganic N yield
Biosolids 1x rate yield
Biosolids 2x rate yield
Inorganic N protein
Biosolids 1x rate protein
Biosolids 2x rate protein

(b)

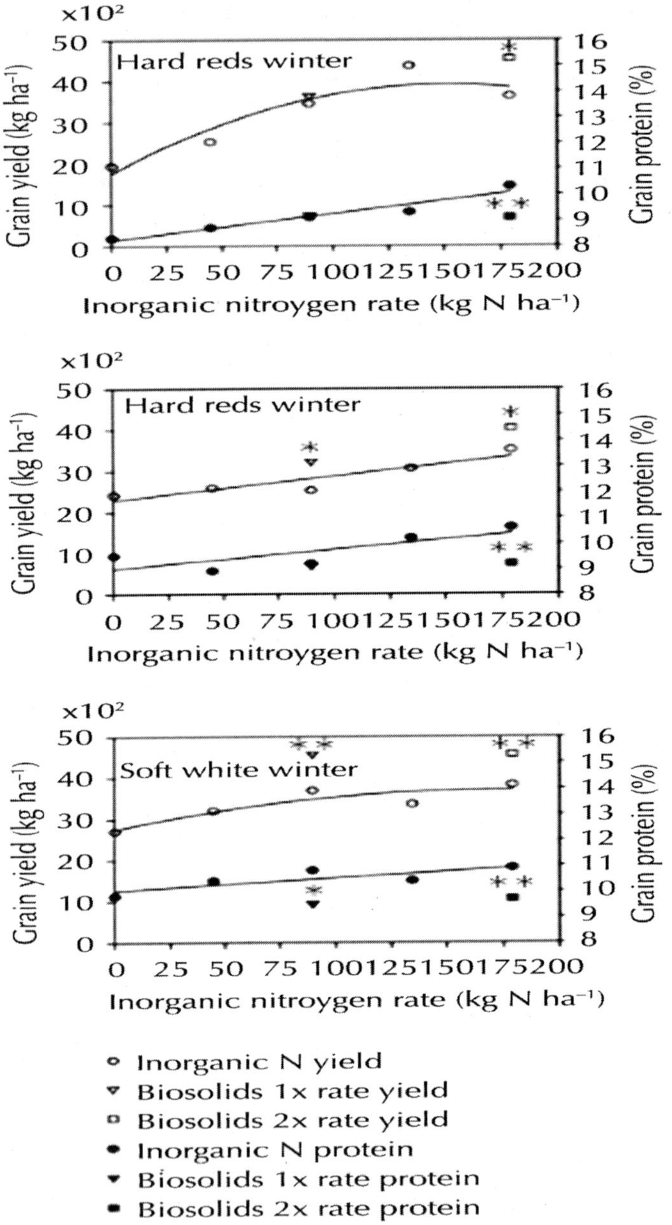


(c)

**Figure 2:** The effect of inorganic N fertilizer and biosolids on yield and grain protein content of three wheat market classes at Lind (a), Davenport (b), and

Pullman (c), Wash in 2008. The horizontal axis (N rate) is the same for Lind and Davenport but higher for Pullman. Due to differences in grain yield, the vertical axis is different for Lind relative to Davenport and Pullman. Where indicated by the regression line, quadratic or linear responses are significant at. If no regression line is included the relationship was not significant (). * And ** indicate significant differences at the 0.10 and 0.05 level, respectively, between inorganic N and biosolids treatments at that inorganic nitrogen and biosolids rate.

Three market classes of winter wheat (soft white, hard white, hard red) were planted at each location. For soft white, cv. "Eltan" was planted at Lind and Davenport and "Madsen" planted at Pullman in 2006. The same cultivars were planted in 2007 but Madsen was used at both Davenport and Pullman. The hard red cv. "Bauermeister" was planted at Lind, and "Paladin" was planted at Davenport and Pullman for both years of the study. The hard white cv. "MDM" was planted at all three sites and for both years. Seeding rates were 50 kg $ha^{-1}$at Lind and 90 kg $ha^{-1}$ at Davenport and Pullman. Row spacing was 19 cm at Davenport and Pullman and 40 cm at Lind.

Individual plot dimensions were 2.1 (Davenport and Pullman) or 2.5 (Lind) m wide by 15.4 m long. A 1.5 m wide by 12.3 m long (18.5 $m^2$) area was harvested with a small combine from the center of each plot at physiological maturity. Grain density (test weight) was measured using standard procedures. Grain protein content was measured using near infrared spectroscopy. Postharvest soil samples were collected in the 0, mid, and high fertilizer N rate treatments as well as the 1× and 2× biosolids treatments of the hard red winter wheat market class only. A composite sample of three cores was collected from each plot in 0.3-m increments to a depth of 1.5 m. Postharvest soil samples were analyzed for residual nitrate (0 to 1.5 m depth) and ammonium (surface 0.3 m only).

## Statistics

Wheat market class was treated as the main plot and fertility treatment the subplot in a randomized complete block, split-plot design with three replications at each location. There were significant site by year and site by N rate interactions within years. Therefore, data were analyzed and presented separately by site and year. Yield and grain protein responses to fertilizer N rate were evaluated using first (linear) and second (quadratic) order polynomial models (Figures 1 and 2). Biosolids at 1× and 2× rates were compared at the intermediate and high inorganic N rates, respectively, using ANOVA. Postharvest residual soil N was analyzed using ANOVA with soil depth and treatment variables. Least significant difference (LSD) values at the 5% level were calculated and presented graphically as bars on the soil profile N graphs (Figure 3). Treatment effects on the soil inorganic N balance were separated using Tukey's Honest Significant Difference at $\alpha=0.05$ (Figure 4).

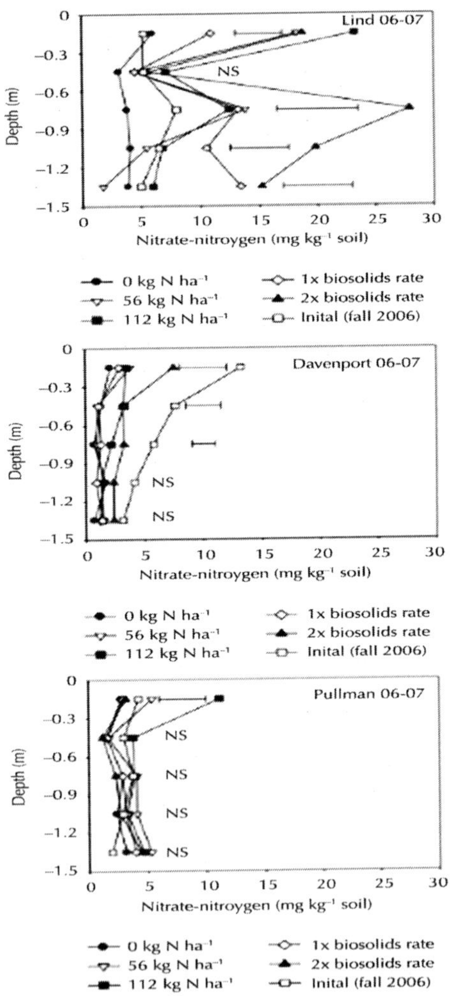


(a)

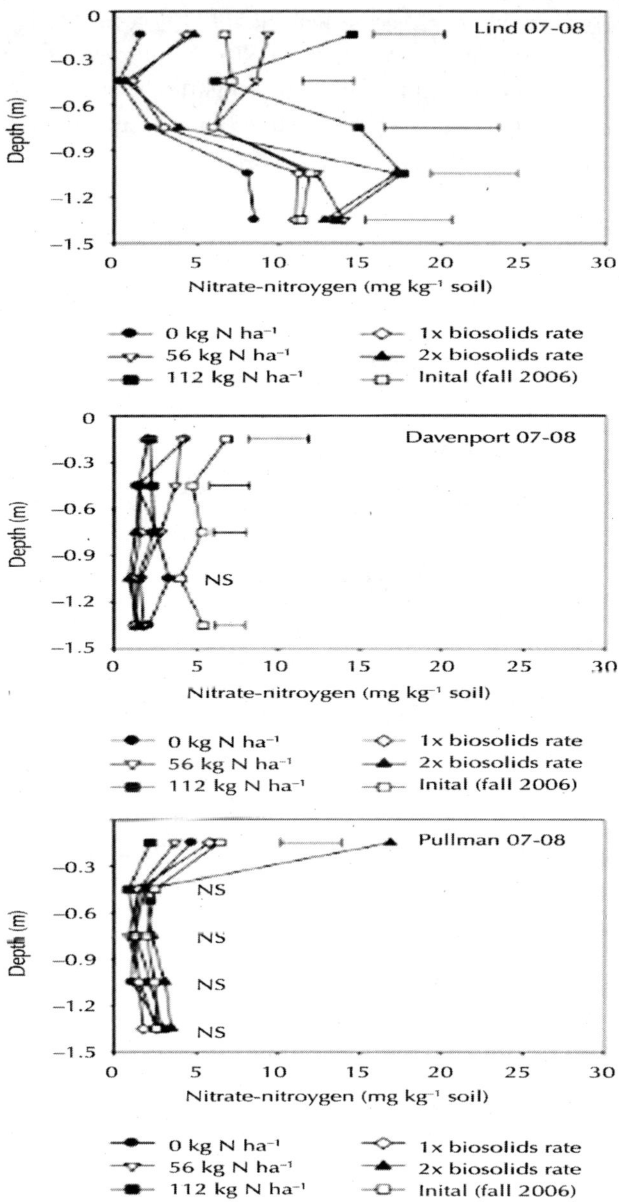


(b)

**Figure 3:** Postharvest soil profile N ($NH_4$ + $NO_3$ for 0 to 0.3 m depth + $NO_3$ only 0.3 to 1.5 m depth) for the hard red winter wheat treatments from the

Lind, Davenport, and Pullman study locations in 2006-2007 (a) and 2007-2008 (b). Error bars represent Least Significant Difference (LSD) values at the 5% level. NS: no significant difference in $NO_3$-N concentration among treatments for that sampling depth. Note that inorganic N rates were the same for all site-years except Pullman 2007-2008, where N rates were higher. This is reflected in the different legend for that panel.

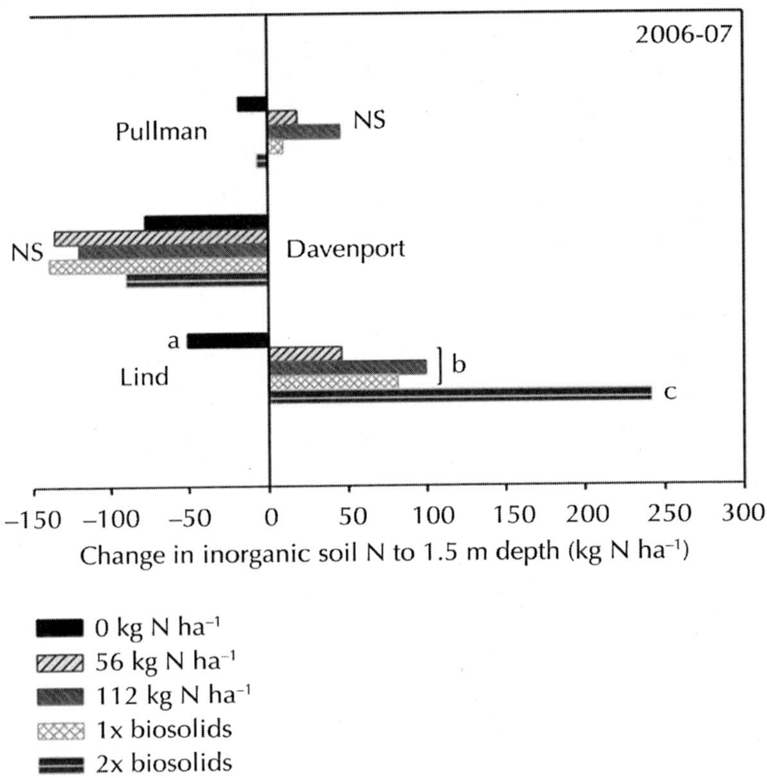


(a)

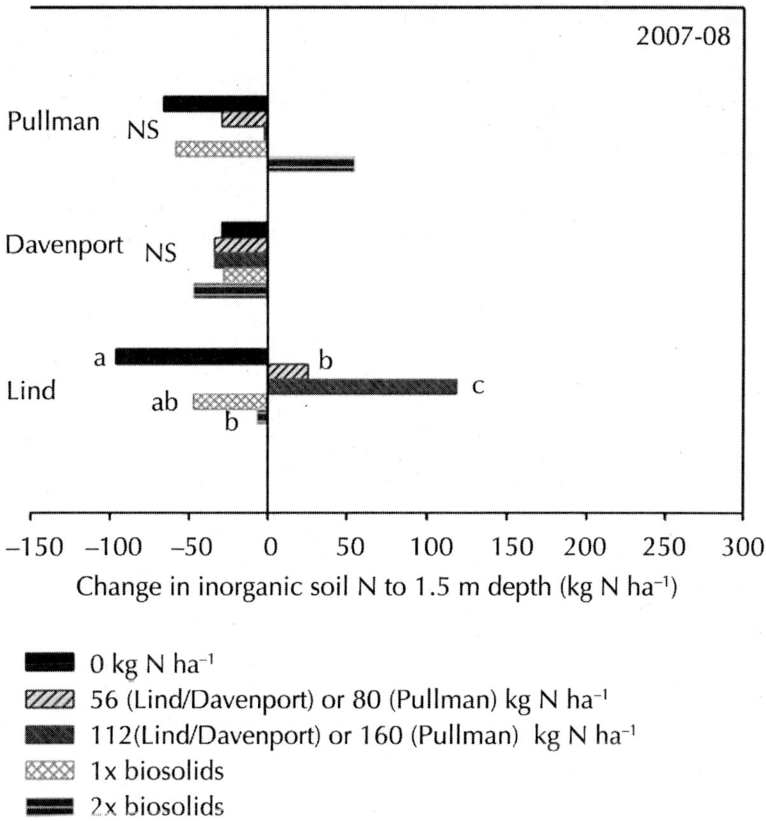


(b)

**Figure 4:** Net changes in total inorganic soil N in a 1.5-m profile between the preplant (Table 1) and postharvest soil sampling of selected fertility treatments in hard red winter wheat at the Lind, Davenport, and Pullman study locations in 2006-07 (a) and 2007-08 (b). Values in the positive (right of vertical line) represent a net gain in soil profile inorganic N compared to the initial (preplant) sampling while values in the negative (left of the vertical line) represent a new depletion of soil profile N. Note the differences in inorganic N application rate between Lind/Davenport and Pullman in 2007-08.

# RESULTS AND DISCUSSION

## Precipitation and Initial Soil Properties at Each Location

Precipitation was near the long-term averages for each location during the 2006-2007 and 2007-2008 crop years (Table 1). Residual plant-available soil moisture generally reflected previous management and precipitation received at each site, where a year of fallow preceding the crop year (Lind and Davenport) was reflected in soil profile moisture measured prior to planting. There was significant residual N at the Lind and Davenport sites. This is largely due to local practice and cropping history, where soil testing is rarely undertaken in low rainfall/low yielding sites like Lind, and a year of fallow preceding the winter wheat crop permits additional organic N mineralization leading to higher preplant residual N levels. Soil test phosphorus (P) was below the recommended critical value for two of the site-years (Lind and Davenport 2006-2007) and near the critical value for Lind 2007-2008. Again, this is a reflection of local practice where, due to low yields, farmers rarely use P fertilizer in the Lind area and only occasionally use P at Davenport, relative to Pullman. Soil test sulfur (S) was below critical values at all site-years.

## Wheat Response to Inorganic N Fertilizer and Biosolids

Grain yield responses to inorganic N were exclusively quadratic in 2007 (Figure 1) and quadratic for three of the nine site × market class datasets in 2008 (Figure 2). At Davenport, the high N rate-induced depression of grain yields in 2007 was associated with wheat lodging. Lodging was not a problem at Lind or Pullman in 2007. In 2008, grain test weight at these locations (data not presented) tended to be lower at high N rates indicating stress caused, perhaps, by moisture depletion. In 2008, grain yield did not respond to inorganic N at Lind (Figure 2). Moisture limitation at Lind, as evidenced by the low 2007-2008 crop year precipitation and preplant residual soil moisture (Table 1), coupled with the relatively high residual N in this year, apparently limited yield responses to N. Precipitation and preplant soil moisture at Pullman were higher during the 2007-08 crop year and would explain the lack of negative response to high N rate at this location.

Quadratic responses to inorganic N fertilizer are common in dryland wheat fertility experiments conducted in this Mediterranean environment [3, 12], but not as dramatic in environments where more rainfall occurs during the summer growing season and biosolids were used as the nutrient source [8, 13]. Negative yield responses to N have been documented in Australia and are generally referred to as "haying off" [14]. The explanation of this phenomenon is that high levels of available N induce a flush of vegetative growth that depletes moisture early in the growing season. This leads to a postanthesis water deficit and severely restricts the translocation of pre-anthesis carbon to the grain, as well as postanthesis assimilation of new carbon destined for wheat kernels.

Grain protein concentration increased linearly with fertilizer N rate for all market classes and locations in 2007 (Figure 1) and at the Davenport and Pullman locations in 2008 (Figure 2). A linear response of grain protein concentration to N rate is common in N response studies of cereals [12], even when yield plateaus or is negatively impacted by high N rates [15]. Soft white winter wheat grain proteins were higher than desired across all N rates at Lind. For hard red and white market classes, yield and grain protein concentration response to fertilizer N suggest fertilizing at rates above maximum yield, and consequently reducing grain yield, may be required to achieve target protein concentrations for each market class. This is likely an issue of timing of fertilization and positional availability of N since previous research has shown that plant-available N residing deep in the profile is more effective than shallow or recently applied N at contributing to high grain protein concentration in dryland wheat [16, 17].

In both 2007 and 2008, grain yields were frequently higher for biosolids than for inorganic N treatments (Figures 1 and 2). Across site-years, yields with biosolids were 0 to 47% higher than the highest yields achieved with inorganic fertilizer. At the site with the largest differential between fertility sources (2007 Davenport hard red), biosolids produced 1400 kg ha$^{-1}$ (>20%) higher grain yield than the highest fertilizer N yield. In no situation did biosolids result in significantly lower grain yields than an inorganic N treatment.

Initial attempts to target biosolids agronomic rates to match the intermediate (56 or 90 kg N $ha^{-1}$) or high (112 or 180 kg N $ha^{-1}$) N fertilizer rates were apparently unsuccessful based on N availability calculations as a fraction of total N in the materials (Table 2). These availability indices were developed from studies in which biosolids were applied in the fallow period of a crop-fallow rotation and were incorporated with tillage [3, 18] or broadcast on the surface for a perennial forage crop [11]. In the present study, biosolids were applied shortly before planting and left on the surface. Both the timing of application relative to sowing and the absence of incorporation would contribute to lower N availability to winter wheat than in previous studies [3]. Evidence from soil profile N measured postharvest (discussed later) also suggests that "Soundgro" biosolids N availability was overestimated in Table 2 calculations. Since the actual N available from biosolids is unknown and can only be estimated, we retained the statistical comparisons between 1× and 2× biosolids treatments and the intermediate and high fertilizer N rates (Figures 1 and 2). While the validity of this comparison could be argued, grain yields and grain protein concentrations with biosolids compared to fertilizer N were substantially higher and lower, respectively, across fertilizer N rates, and results of statistical comparisons would be similar regardless of the inorganic N rate selected for comparison to biosolids.

Grain yield responses to biosolids did not reflect a "haying off" effect observed with fertilizer N (discussed above). One explanation for this is that the rate of N mineralization from biosolids was not rapid enough to stimulate a flush of vegetative growth and lead to moisture depletion and stress. In fact, grain test weights of biosolids treatments were within acceptable ranges and did not indicate widespread stress. Also, higher yields from biosolids compared to fertilizer N were observed in 2008 when there was little indication of a negative yield response to high rates of fertilizer N. Previous studies comparing biosolids and fertilizer N responses in dryland areas of the PNW did not show consistently higher yields with biosolids materials [3]. A second explanation for the higher yields from biosolids in the present study is that other nutrients were deficient at these sites and were supplied in adequate quantities by the biosolids materials. At Lind and Davenport in 2007, soil test P was below critical levels while S was below = for all site-years (Table 1). While some P and S was applied at as starter fertilizer with the seed at Davenport and Pullman, the rates were relatively low (5 kg P and

13 kg S $ha^{-1}$). The additional P and S associated with the ===60y8.o;h may explain the higher yields associated with these treatments across site years.

In situations where grain yield was higher with biosolids than with comparable inorganic N treatments, grain protein content was often lower. This effect could be explained by the dilution of grain N in these higher yielding treatments, leading to lower grain protein. The inverse relationship between grain yield and grain protein concentration is a common phenomenon in wheat [15]. Another factor in this study may be the positional availability of N in the soil profile. Previous research has shown that high grain protein concentration is more easily achieved when N is available for uptake later in the season during grain filling [16, 17]. In this dryland area where the soil surface dries rapidly, N located at depth in the profile where late-season moisture absorption occurs is more likely to contribute to high grain protein concentration. Central to the premise of this study is that biosolids are better suited to producing high protein hard red and white winter wheat than soft white winter wheat in these dryland environments, as was suggested earlier [3]. This does not appear to be the case under the conditions (time, placement) of biosolids application in this study.

## Soil Profile N and N Balance

Postharvest inorganic N distributions for hard red winter wheat were variable among fertility treatments, particularly at Lind (Figure 3). Except for Lind and limited treatments at the 0 to 0.3-m depth at Pullman, postharvest soil profile N was lower for inorganic and biosolids treatments than the initial soil profile N. Interestingly, as evidenced by the higher amount of $NO_3$-N at depth in the profile, there was some N movement to the 0.6 to 1.2-m depth at the Lind location for the biosolids and high N rate treatments (Figure 3) even though this site received the least amount of precipitation and had some of the lowest preplant soil moisture (Table 1). The Lind site had significant preplant residual N in the profile (Table 1; Figure 3). This, coupled with relatively low yields and high N rates, resulted in some increase in soil N with intermediate and high N rate treatments. Regardless, as evidenced by the lack of statistical difference between treatments and the unfertilized control, there was little N movement below the 1.2 m depth in any treatment except at Lind.

When expressed as a net gain or loss of inorganic N from the profile, there was no difference in net N change in the soil profile among treatments at Davenport and Pullman, suggesting no greater risk of soil $NO_3$accumulation with biosolids than with inorganic fertilizer. There was a net increase in soil N for intermediate and high fertilizer rates and biosolids treatments at Lind in 2007 and for fertilizer treatments at Lind in 2008, compared to the unfertilized control. Accumulated N in the profile could be subject to leaching below the crop root zone in the fallow cycle at this location, particularly since much of this residual N is located at depth in the soil profile.

Overall, at the Davenport and Pullman sites, results indicate efficient use of residual and applied N forms regardless of the source (biosolids versus inorganic N). While this is a promising finding, other studies in which high rates of biosolids were applied and/or applications were made in the fallow year, residual soil N increased, and there was some evidence of leaching loss [3, 8].

# CONCLUSIONS

Biosolids applied within two weeks of planting and without incorporation were an efficient source of nutrients for dryland wheat production across a range of rainfall zones in eastern Washington. In situations of low to moderate preplant soil profile N, organic N released by biosolids was well utilized by the crop with residual soil levels no greater than with inorganic N sources. Yields with biosolids were frequently higher than with inorganic N treatments, likely due to a combination of slow N release and contributions of other nutrients such as P and S that were deficient in these systems. When applied in this manner, biosolids were not effective at producing high grain protein concentrations in hard red or hard white winter wheat. If biosolids material was applied in the fallow year in low and intermediate precipitation zones, the impact on grain protein concentration may be greater. Early application of biosolids in the higher precipitation areas where annual cropping is practiced is not possible.

## ACKNOWLEDGMENTS

The authors would like to thank the Northwest Biosolids Management Association for financial support of this research. This research was supported by the Washington State University Agricultural Research Center under Hatch Project no. 0579.


## REFERENCES


1. K. A. Barbarick, J. A. Ippolito, and D. G. Westfall, "Distribution and mineralization of biosolids nitrogen applied to dryland wheat," Journal of Environmental Quality, vol. 25, no. 4, pp. 796–801, 1996
2. C. G. Cogger, T. A. Forge, and G. H. Neilsen, "Biosolids recycling: nitrogen management and soil ecology," Canadian Journal of Soil Science, vol. 86, no. 4, pp. 613–620, 2006
3. D. M. Sullivan, A. I. Bary, C. G. Cogger, and T. E. Shearin, "Predicting biosolids application rates for dryland wheat across a range of Northwest climate zones," Communications in Soil Science and Plant Analysis, vol. 40, no. 11-12, pp. 1770–1789, 2009
4. K. A. Barbarick, J. A. Ippolito, and D. G. Westfall, "Biosolids effect on phosphorus, copper, zinc, nickel, and molybdenum concentrations in dryland wheat," Journal of Environmental Quality, vol. 24, no. 4, pp. 608–611, 1995
5. C. G. Cogger, D. M. Sullivan, A. I. Bary, and J. A. Kropf, "Matching plant-available nitrogen from biosolids with dryland wheat needs," Journal of Production Agriculture, vol. 11, no. 1, pp. 41–47, 1998.
6. D. K. McCool, D. R. Huggins, K. E. Saxton, and A. C. Kennedy, "Factors affecting agricultural sustainability in the Pacific Northwest, USA: an overview," in Proceedings of the 10th International Soil Conservation Organization Meeting on Sustaining the Global Farm Symposium, D. E. Stott, R. H. Mohtar, and G.C. Steinhardt, Eds., pp. 255–260, Purdue University, May 1999.

7. A. C. S. Rao, J. L. Smith, V. K. Jandhyala, R. I. Papendick, and J. F. Parr, "Cultivar and climatic effects on the protein content of soft white winter wheat," Agronomy Journal, vol. 85, no. 5, pp. 1023–1028, 1993.
8. K. A. Barbarick, J. A. Ippolito, and J. McDaniel, "Fifteen years of wheat yield, N uptake, and soil nitrate-N dynamics in a biosolids-amended agroecosystem," Agriculture, Ecosystems and Environment, vol. 139, no. 1-2, pp. 116–120, 2010
9. R. Koenig, Eastern Washington Nutrient Management Guide: Dryland Winter Wheat, Washington State University Extension Bulletin, no. 1987, Washington State University Extension, Pullman, Wash, USA, 2005,
10. J. T. Gilmour, C. G. Cogger, L. W. Jacobs, G. K. Evanylo, and D. M. Sullivan, "Decomposition and plant-available nitrogen in biosolids: laboratory studies, field studies, and computer simulation," Journal of Environmental Quality, vol. 32, no. 4, pp. 1498–1507, 2003.
11. C. G. Cogger, A. I. Bary, and E. A. Myhre, "Estimating nitrogen availability of heat-dried biosolids,"Applied and Environmental Soil Science. In press.
12. S. O. Guy and R. M. Gareau, "Crop rotation, residue durability, and nitrogen fertilizer effects on winter wheat production," Journal of Production Agriculture, vol. 11, no. 4, pp. 457–461, 1998.
13. K. A. Barbarick and J. A. Ippolito, "Nutrient assessment of a dryland wheat agroecosystem after 12 years of biosolids applications," Agronomy Journal, vol. 99, no. 3, pp. 715–722, 2007
14. A. F. van Herwaarden, G. D. Farquhar, J. F. Angus, R. A. Richard, and G. N. Howe, "'Haying-off', the negative grain yield response of dryland wheat to nitrogen fertiliser. I. Biomass, grain yield, and water use," Australian Journal of Agricultural Research, vol. 49, no. 7, pp. 1067–1081, 1998.
15. G. L. Terman, R. E. Ramig, A. F. Dreier, and R. A. Olson, "Yield-protein relationships in wheat grain as affected by nitrogen and water," Agronomy Journal, vol. 61, pp. 755–759, 1969.
16. V. L. Cochran, R. L. Warner, and R. I. Papendick, "Effect of N depth and application rate on yield, protein content and quality of winter wheat," Agronomy Journal, vol. 70, pp. 964–968, 1978.

17. K. E. Sowers, B. C. Miller, and W. L. Pan, "Optimizing yield and grain protein in soft white winter wheat with split nitrogen applications," Agronomy Journal, vol. 86, no. 6, pp. 1020–1025, 1994.

18. C. G. Cogger and D. M. Sullivan, Worksheet for Calculating Biosolids Application Rates in Agriculture, Pacific Northwest Extension Bulletin, no. 511, Washington State University Extension, Pullman, Wash, USA, 2007.

Chapter 6

# Spectral Indices: in-Season Dry Mass and Yield Relationship of Flue-cured Tobacco under Different Planting Dates and Fertiliser Levels

Ezekia Svotwa[1], J. Anxious Masuka[2], Barbara Maasdorp[3], and Amon Murwira[3]

[1]Department of Crop Science, University of Zimbabwe, Harare, Zimbabwe

[2]Department of Geography and Environmental Studies, University of Zimbabwe, Harare, Zimbabwe

[3]Tobacco Research Board/Kutsaga Research Station, Harare, Zimbabwe

# ABSTRACT

This experiment investigated the relationship between tobacco canopy spectral characteristics and tobacco biomass. A completely randomized design, with plantings on the 15th of September, October, November, and December, each with 9 variety × fertiliser management treatments, was used. Starting from 6 weeks after planting, reflectance measurements were taken from one row, using a multispectral radiometer. Individual plants from the other 3 rows were also measured, and the above ground whole plants were harvested and dried for reflectance/dry mass regression analysis. The central row was harvested, cured, and weighed. Both the maximum NDVI and mass at untying declined with later planting and so was the mass-NDVI coefficient of determination. The best fitting curves for the yield-NDVI correlations were quadratic. September reflectance values from the October crop reflectance were statistically similar ($P < 0.05$), while those for the November and the December crops were significantly different ($P < 0.05$) from the former two. Mass at untying and NDVI showed a quadratic relationship in all the three tested varieties. The optimum stage for collecting spectral data for tobacco yield estimation was the 8–12 weeks after planting. The results could be useful in accurate monitoring of crop development patterns for yield forecasting purposes.

# BACKGROUND

Crop yield estimation in many countries is based on conventional techniques of data collection and ground-based field reports [1]. A variety of mathematical models relating to crop yield have also been proposed in recent years for many crops [2, 3]. In Zimbabwe crop surveys are mostly used in estimating crop yield [4]. The method is costly, time consuming, and is prone to large errors due to incomplete ground observations, leading to poor crop yield assessment and crop area estimations [1].

Remote sensing data has the potential and the capacity to provide spatial information in a global scale of features and phenomena on earth on an almost real-time basis [1]. Use of remote sensing techniques has the potential to provide quantitative and timely information on agricultural crops over large areas, and many different methods have been developed to estimate crop yields [5–7]. In general, the use of remote sensing was aimed at reducing the number of samples of ground surveys, making it less expensive [8]. With the application of remote sensing in agriculture, there is potential not only in identifying crop classes but also in estimating crop yield [1].

Spectral measurements from crops can be used in estimating crop parameters such as leaf area index [9], plant population, and even canopy total nitrogen status during the growth cycle of the crop [10]. Vegetation indices are algorithms aimed at simplifying data from multiple reflectance bands to a single value correlating with physical vegetation parameters, such as biomass, productivity, leaf area index, or percent vegetation ground cover [11]. Single band reflectance is combined into a vegetation index in order to minimize the effect of such factors as optical properties of the soil background and illumination and view geometric as well as meteorological factors on the canopy radiometric properties [12].

Vegetation indices, as summarized by Gross [13], are based on the characteristic reflection of plant leaves in the visible and near-infrared portions of light. By applying a "vegetation index" to the satellite imagery, concentration of green leaf vegetation can be quantified [14]. As explained by Liew [15] healthy vegetation has low reflection of visible light (from 0.4 to 0.7 μm), since the visible light is strongly absorbed by chlorophyll for photosynthesis and, at the same time, there is high reflection of near-infrared light (from 0.7 to 1.1 μm). The portion of reflected near-infrared light depends on the cell structure of the leaf [16]. In fading or unhealthy leaves, photosynthesis decreases and cell structure collapses resulting in an increase of reflected visible light and a decrease of reflected near-infrared light [13].

The normalized difference vegetation index (NDVI) has been considered to be a useful way for crop yield assessment models, using various approaches such as simple integration, to reflect vegetation greenness [17]. The index responds to changes in the amount of green biomass, chlorophyll content, and canopy water stress and, hence, is most commonly used in assessing crop vigor, vegetation cover, and biomass production from multispectral satellite data [18–20]. The NDVI is calculated from the near-infrared (NIR) and red (R) bands of either handheld or satellite sensors using the formula NDVI = (NIR − Red)/ (NIR + Red).

The validity of crop yield models with NDVI is determined by the strength of association between the two variables included in the model [21]. It is also essential to have an understanding of the correlation existing between yield and NDVI at different phonological stages of crop for selecting appropriate date of satellite pass to include in the model [22].

Tobacco crop plays an important role in the economy of Zimbabwe, and in the 2012/2013 marketing season, 144 million kg of tobacco was sold, earning the country $525 million [23]. Crop area and yield forecasts play an important role in stabilizing tobacco prices at the auction floors. Crop forecasting is the art of predicting crop yields and production before the harvest actually takes place, typically a couple of months in advance. Zimbabwe mostly relies on crop statistical forecasting/estimation, crop reports/field visits from extension officers, and statistical crop forecasts for crop yield forecasts [4]. However, data from crop estimates, which are obtained through surveys conducted after harvests, are in most countries available quite lately for early warning purposes.

An overestimation of the crop would jeopardize the grower's profit in that it causes fall in prices when supply exceeds the estimated volume. Underestimation, on the other hand, causes unnecessary panic and competition among buyers of the crop, causing a rise in the price of the crop. The timely evaluation of potential crop yields in general becomes important because of the huge economic impact crops have on the world markets [5] and in particular on the economy of Zimbabwe.

Remotely sensed measurements can be used in monitoring the effects of agronomic practices, which are considered in developing yield prediction models [24]. A more direct remote sensing data yield, described in simple formulae, without deeper physiological background, is simpler to use and easier to understand [25] and would be applicable in tobacco, where the target, the leaf, is the harvestable part. This experiment investigated the relationship between canopy spectral characteristics of three tobacco varieties established on three planting dates and, under three fertilizer regimes, in-season dry matter and final yield. It was assumed in this study that the most suitable stage to predict yield is that where the canopy NDVI was most positively correlated with in season dry mass, and a model relating the NDVI for this stage to cured leaf mass would be established. It was also hypothesized that the strength of the relationship between in-season dry mass and yields expressed as mass at untying with NDVI is not affected by tobacco variety, planting date, and fertiliser application rate. The results for the project will be used to select the most appropriate stage of collecting remote sensing data for field level and national tobacco crop area and yield forecasting. This information could be very useful in relating the reflectance measured from the tobacco cropped lands to in-season crop condition and final yield and quality predictions using remote sensing.

# METHOD

Study Area. The experiment was conducted at Kutsaga Research Station in Zimbabwe in the 2010–2012 cropping seasons. Kutsaga is located between longitude 31° 08′E, latitude 17° 55′S, and at an altitude of 1000 m to 1500 m [10]. The long-term annual average rainfall is 850 mm.

The experimental plots were located on well-drained granitic sands. During February of 2009 and 2010 the plots were disked after a three-year Katambora grass fallow period to incorporate grass. Agricultural lime was applied using recommendations given from soil test results to raise the soil pH from 5.3 to 6.3 optimum for tobacco production. For the three years preceding the 2008 experiments, the sites were under Katambora grass to control nematodes. Recommended cultural and management practices were followed [10], except regard N: P: K levels and planting times, which were treatments in the experiment.

## Fertilizer Treatments

In order to establish the relationship between spectral data and yield, there was a need to create variable growth conditions [26], and three varieties, four planting dates, and three fertilizer levels were tested. The variety-fertilizer treatments were applied by hands as shown in Table 1. The N:P:K treatment was hand-applied in bands about 10 cm deep and 30 cm to each side of a row at planting, while N treatments were applied at about 4 weeks after transplanting and after topping (at 6 weeks after planting).

**Table 1:** Variety-fertilizer treatments

| Treatment | Description |
|---|---|
| (1) | K RK 26—50% recommended fertiliser |
| (2) | K RK 26—recommended fertiliser |
| (3) | K RK 26—150% recommended fertiliser |
| (4) | T 66—50% recommended fertiliser |
| (5) | T 66—recommended fertiliser |
| (6) | T 66—150% recommended fertiliser |
| (7) | K E1—50% recommended fertiliser |
| (8) | K E1—recommended fertiliser |
| (9) | K E1—150% recommended fertiliser |

The experiment was laid out in a completely randomized design with plantings on September 15, October 15, November 15, and December 15 each with 9 variety × fertilizer management treatments (Table 1). Three tobacco varieties K RK26, T 66, and K E1, developed by Kutsaga Research Station, were used, while three fertiliser management levels (50%, 100%, and 150% recommended) were applied by hand (Table 1). The recommended compound fertiliser rate from soil test results was 700 kg/ha, while that for ammonium nitrate (34.5% N) was 96 kg/ha at 4 weeks after planting and 75 kg/ha after topping.

## Procedure

Radiometric measurements were made weekly from the age of 6 weeks after planting on 5 m × 5 m square sampling plots, using a handheld multispectral radiometer (Cropscan MSR-5, 450–1750 nm), with the FOV centering over rows. All treatment applications had been completed at this stage of development.

Spectral data from the 4 planting dates were used to construct temporal NDVI profiles and one, with uninterrupted growth was selected for the in-season dry-mass-NDVI regression analysis. Above ground samples were collected after a corresponding canopy reflectance measurement had been obtained. Some 10 plants were sampled from each variety × fertiliser × planting time treatment after spectral data collection at 8, 10, 12, and 14 weeks after planting. The data collection timing, during midday and under cloudless conditions across the growing season, could include bare soil, early crop growth stage, peak crop greenness, and crop maturity imagery [27].

Each sampling plot measured consisted 5 rows of each with 32 plants spaced at 56 cm. The interrow distance was 1.2 m. normalized difference vegetation indices (NDVI) was calculated from the spectral bands obtained in Channels 3 and 4 of the MSR 5 which correspond to the visible (VIS) and near-infra red (NIR), respectively, using the following formula:

$$\text{NDVI} = \frac{\text{nir} - \text{red}}{\text{nir} + \text{red}}. \qquad (1)$$

The multispectral radiometer (MRS 5) was positioned facing vertically downward at 1 m above crop canopies, and measurements were taken around solar noon to minimize the effect of diurnal changes in solar zenith angle. In total, 10 measurements were taken per sampling area and reflectance measurements were then averaged for each sampling plot to estimate a single reflectance value. Mature leaves were harvested from one row, cured, and yield determined before handle losses during crop grading as applied by Zhang et al. [28].

Reflectance measurements were also taken on individual plants from the other 3 rows. After taking reflectance readings, the above ground whole plants were harvested and packed in khaki bags and dried in microbarns. Dry matter measurements were later taken for reflectance/DM regression analysis. Three rows were also harvested and cured and mass was determined just after curing, before handling losses were incurred. The NDVI for the growth stage where there is the highest in-season dry-mass-NDVI correlation was selected for determining the mass-at-untying NDVI correlations.

Three-dimensional positions, latitude, longitude, and altitude, for the whole experimental area and for each treatment plot will be taken using a Garmin personal navigator (GPS V) to enable repeated sampling at the same location. Yield data were collected at harvest.

## Data Analysis

NDVI data was analysed by analysis of variance and statistically significant treatment effects were separated using least significant differences (LSDs). The data was analysed using the Genstat 9.2 statistical package at 5% level of significance. Student's t-test calculations were done to compare the planting date effect, and graphs were plotted using Excel 2007.

# RESULTS

The NDVI for all four planting date treatments crop rose from week 6 after planting to peak from 9 to 12 weeks after planting (Figure 1). At eight weeks after planting all the variety × fertiliser treatments in all, except for the December crop, started showing significant ($P < 0.05$) treatment differences. Beyond the peak, the NDVI also fell gently to reach the minimum at 13 weeks of age. The October crop (Figure 1(b)) was sampled for NDVI versus in-season dry mass analysis. This crop was selected because it was not subjected to long dry conditions after planting and had a good establishment. In addition temporal NDVI profile for the planting date had the highest NDVI value, which would enable a wide variation of the DM-NDVI relationships (Figure 1(b)).

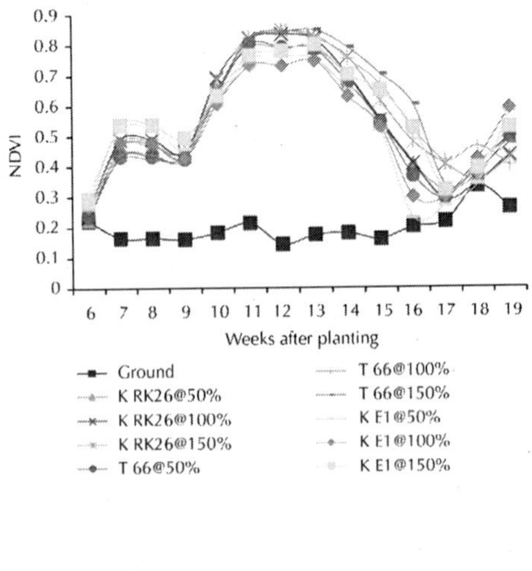


(a)

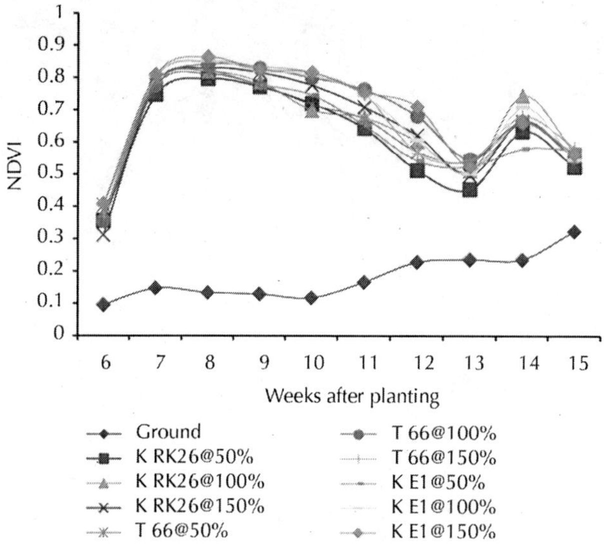
NDVI
Weeks after planting
Ground
K RK26@50%
K RK26@100%
K RK26@150%
T 66@50%
T 66@100%
T 66@150%
K E1@50%
K E1@100%
K E1@150%

(b)

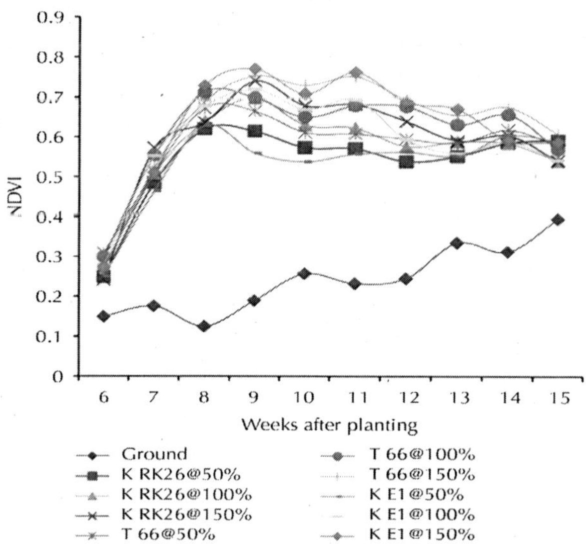
NDVI
Weeks after planting
Ground
K RK26@50%
K RK26@100%
K RK26@150%
T 66@50%
T 66@100%
T 66@150%
K E1@50%
K E1@100%
K E1@150%

(c)

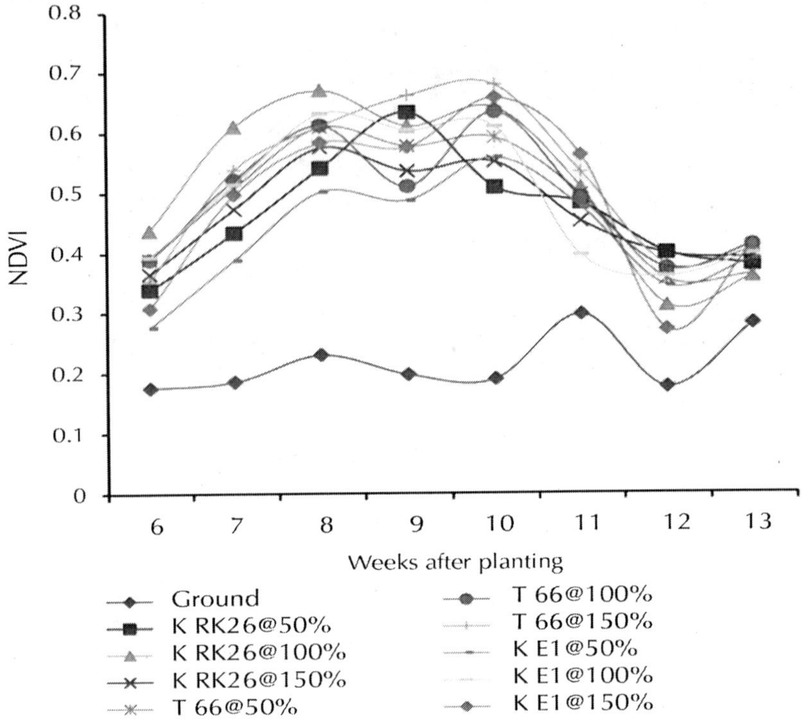


(d)

**Figure 1:** The NDVI temporal profiles for the (a) September, (b) October, (c) November, and (d) December planted crops.

The correlation between NDVI and in-season dry mass became stronger from the first sampling date (8 weeks after planting), reaching the highest at 10 weeks and later declined (Figure 2). Plants were not sampled after week 14 because reaping had become intense and some plants had already been stripped.

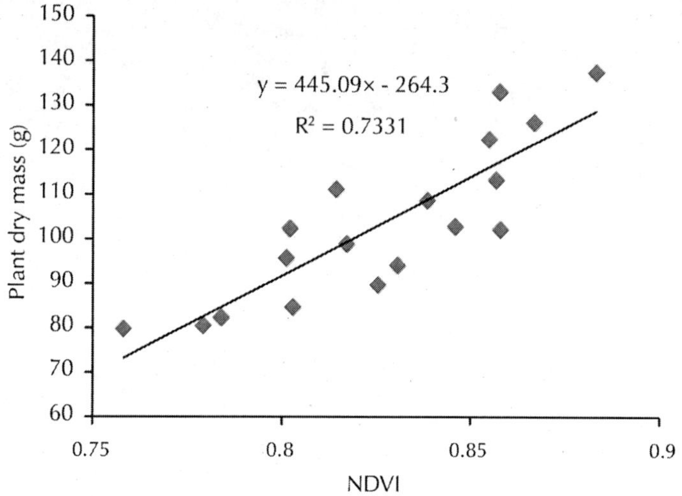
150
140
130
120
110
100
90
80
70
60
Plant dry mass (g)
y = 445.09x - 264.3
R² = 0.7331
0.75
0.8
0.85
0.9
NDVI

(a)

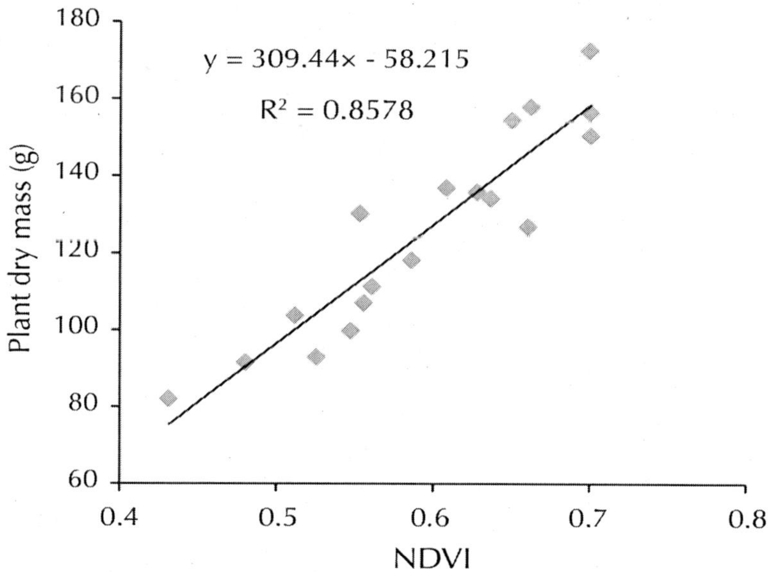
180
160
140
120
100
80
60
Plant dry mass (g)
y = 309.44x - 58.215
R² = 0.8578
0.4
0.5
0.6
0.7
0.8
NDVI

(b)

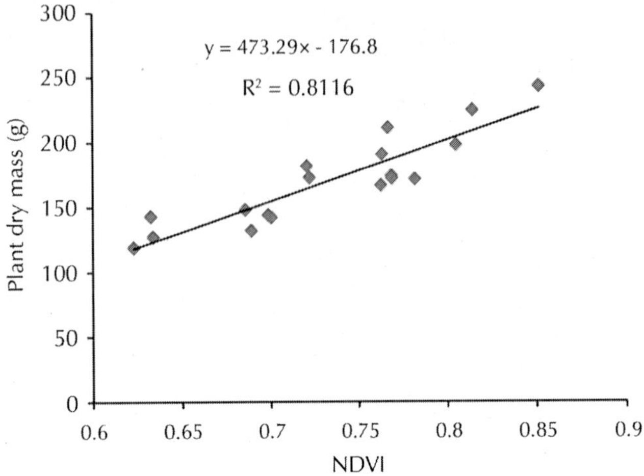


(c)

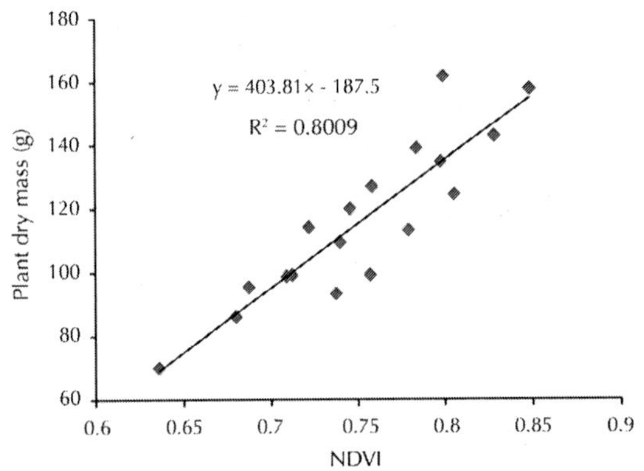


(d)

**Figure 2:** Dry-mass-reflectance correlations for the flue-cured tobacco samples collected from the (a) September, (b) October, (c) November, and (d) December planted crops.

The NDVI response to variety × fertiliser treatment was similar to that for mass at untying (Figure 3). There was a general decline in both maximum NDVI and mass at untying with later planting, with the least values attained in December planting.

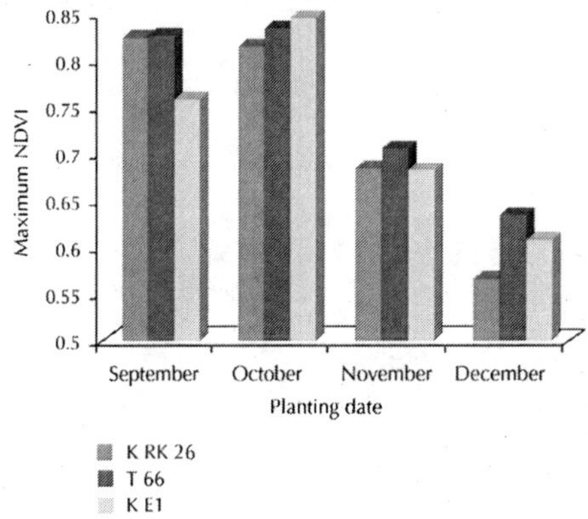


(a)

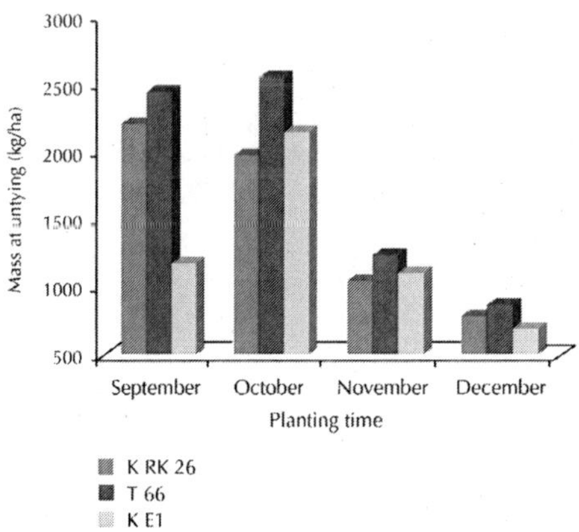


(b)

**Figure 3:** (a) Tobacco varieties' maximum NDVI and (b) mass at untying response to planting date.

The mass at untying-NDVI coefficient of determination also decreased with later planting, with least being that of December (Figure 4). In all the four planting date treatments, the best fitting curves for mass at untying and NDVI correlations were quadratic. The September coefficient of determination ($r^2$ = 0.79) was the highest as compared to October ($r^2$ = 0.594), November ($r^2$ = 0.695), and December ($r^2$ = 0.515).

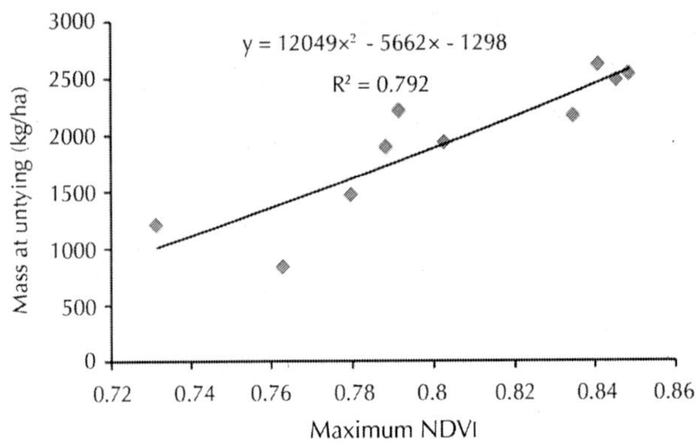


(a)

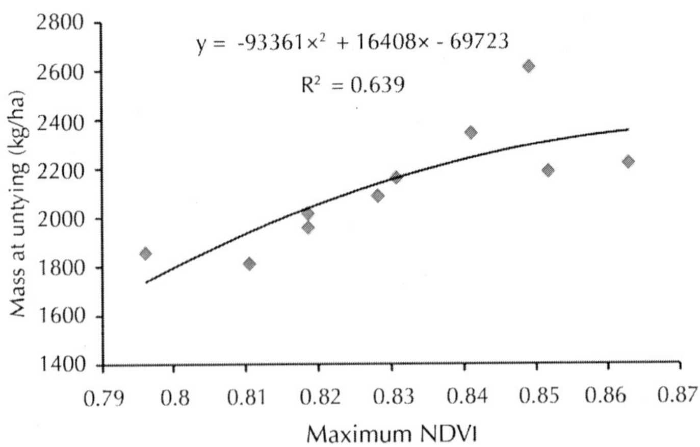


(b)

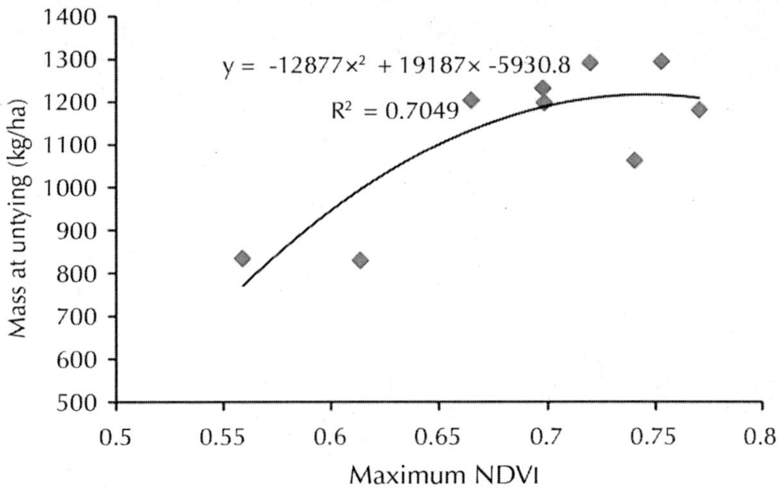


(c)

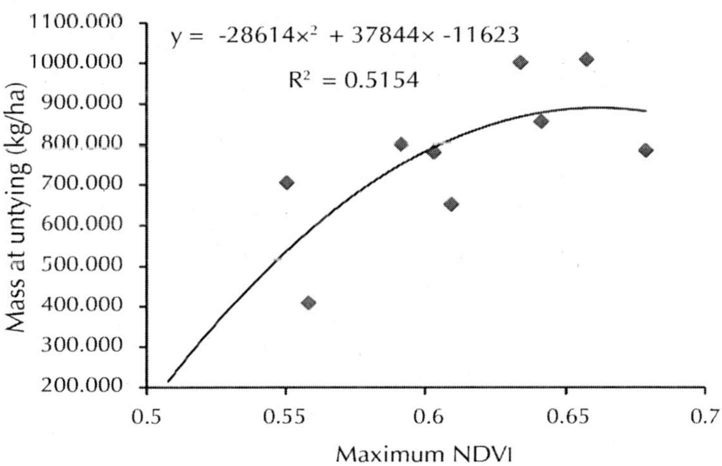


(d)

**Figure 4:** The relationship between the (a) September, (b) October, (c) November, and (d) December planted tobacco mass at untying and maximum NDVI.

The September and the October crop reflectance values were statistically similar (), while the November and the December reflectance values were, for each, significantly different ($P < 0.05$) from all the rest (Table 2).

**Table 2:** t-test for the comparison of mean maximum NDVI values for the different planting date treatments

| | | **P values** | |
|---|---|---|---|
| | **October** | **November** | **December** |
| September | 0.058204 | 0.000494993 | $4.883E-06$ |
| October | | $5.66702E-06$ | $1.08556E-08$ |
| November | | | 0.000152201 |

All the three varieties, K RK 26, T 66, and K E1, showed a quadratic relatioship between mass at untying and NDVI (Figure 5). K E1 had the highest ($r^2 = 0.86$) mass at untying NDVI coefficient of determination. The mass at untying NDVI correlations in K E1 ($r$ 2 = 0.748) and T 66 ($r^2 = 0.773$) were comparable. In all the three varieties, the best fitting curves for mass at untying versus NDVI correlations were also quadratic, with comparable gradients at NDVI values between 0.65 and 0.75 (Table 3).

**Table 3:** Variety yield-NDVI gradients at NDVI = 0.65, 0.7, and 0.75

| **NDVI** | **0.65** | **0.7** | **0.75** |
|---|---|---|---|
| | **Yield-NDVI gradient** | | |
| K RK 26 | 4456.16 | 4776.4 | 5256.76 |
| T66 | 4505.72 | 5448.8 | 6863.42 |
| KE 1 | 4022.96 | 4776.4 | 5906.56 |

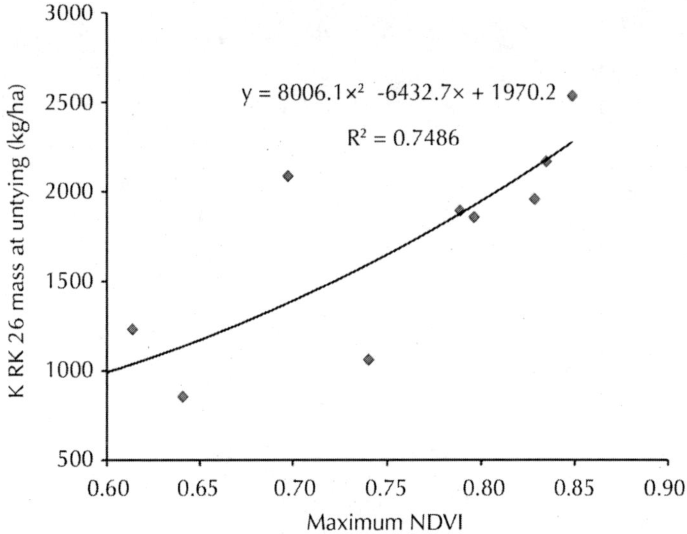
y = 8006.1x² -6432.7x + 1970.2
R² = 0.7486
K RK 26 mass at untying (kg/ha)
3000
2500
2000
1500
1000
500
0.60
0.65
0.70
0.75
0.80
0.85
0.90
Maximum NDVI

(a)

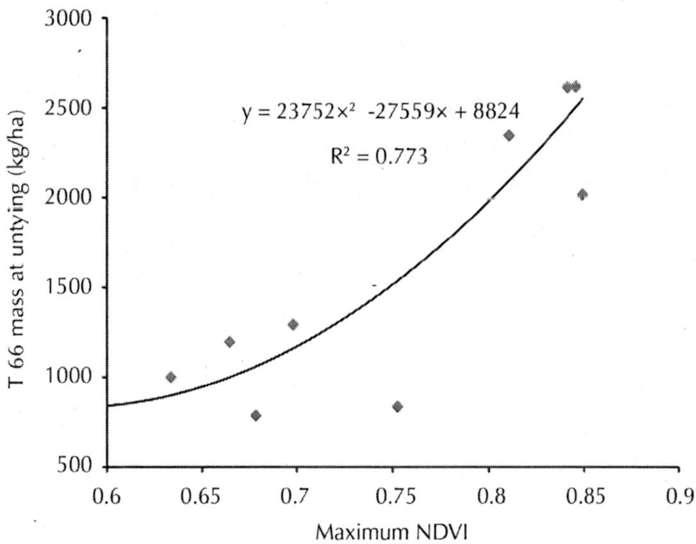
y = 23752x² -27559x + 8824
R² = 0.773
T 66 mass at untying (kg/ha)
3000
2500
2000
1500
1000
500
0.6
0.65
0.7
0.75
0.8
0.85
0.9
Maximum NDVI

(b)

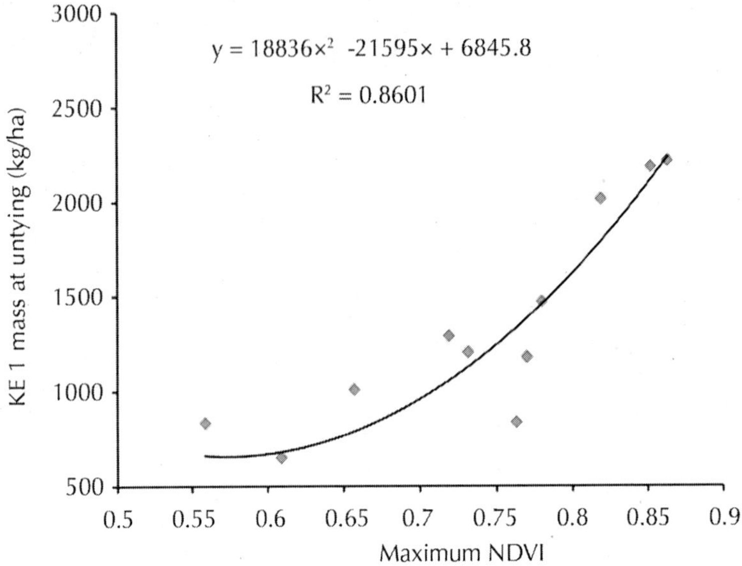


(c)

**Figure 5:** The relationship between tobacco varieties (a) K RK26, (b) T 66, and (c) K E1 mass at untying and maximum NDVI.

All the three fertilizer levels, 50% ($r^2$ = 0.925), 100% ($r^2$ = 0.966), 150% ($r^2$ = 0.92), displayed equally strong mass at untying NDVI relationship (Figure 6) with comparable gradients, again at NDVI value between 0.65 and 0.7 (Table 4).

**Table 4:** Fertiliser level yield-NDVI gradients at NDVI = 0.65, 0.7, and 0.75

| **NDVI** | **0.65** | **0.7** | **0.75** |
|---|---|---|---|
| | **Yield-NDVI gradient** | | |
| 50% fertiliser level | 4360.5 | 5553.0 | 6745.5 |
| 100% fertiliser level | 4288.1 | 5689.8 | 7091.5 |
| 150% fertiliser level | 2260.4 | 5399.2 | 8538.0 |

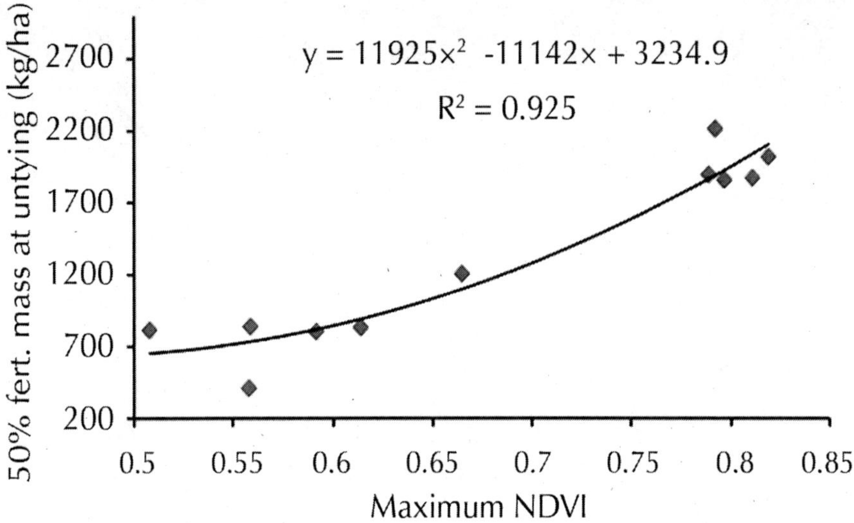
y = 11925x² -11142x + 3234.9
R² = 0.925
50% fert. mass at untying (kg/ha)
2700
2200
1700
1200
700
200
0.5
0.55
0.6
0.65
0.7
0.75
0.8
0.85
Maximum NDVI

(a)

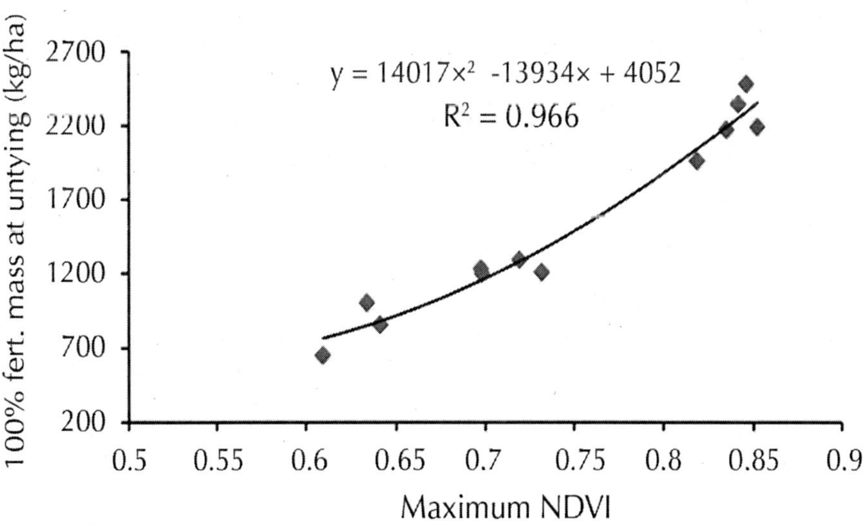
y = 14017x² -13934x + 4052
R² = 0.966
100% fert. mass at untying (kg/ha)
2700
2200
1700
1200
700
200
0.5
0.55
0.6
0.65
0.7
0.75
0.8
0.85
0.9
Maximum NDVI

(b)

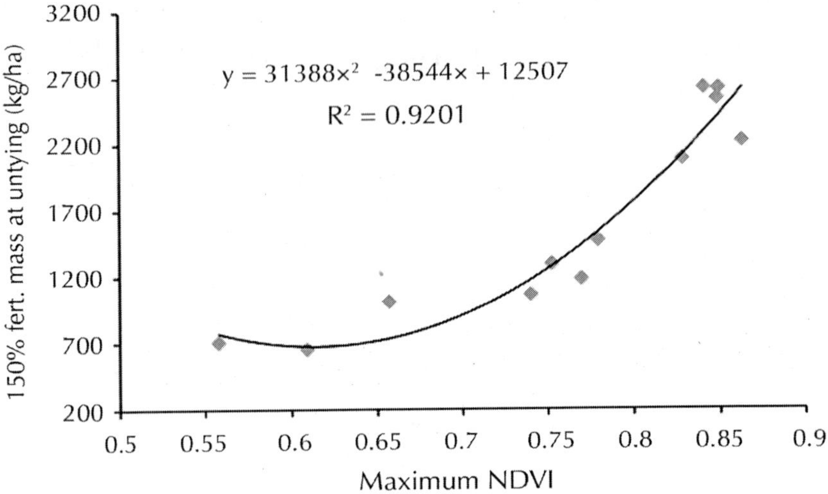


(c)

**Figure 6:** The relationship between tobacco mass at untying for the (a) 50%, (b) 100%, and (c) 150% fertilizer rate treatments and maximum NDVI.

The yield, expressed as mass at untying (kg/ha), for the September-October, November, and December planting can, therefore be estimated separately by the models in Figure 7:

1 $y = -26708x^2 + 54365x - 24516,$

$R^2 = 0.741,$ (2)

2 $y = -12877x^2 + 19187x - 5930,$

$R^2 = 0.704,$ (3)

3 $y = -28614x^2 + 37844x - 11623,$

$R^2 = 0.515,$ (4)

Where $y$ is the mass at untying and $x$ is the maximum normalized difference vegetative index (NDVI).

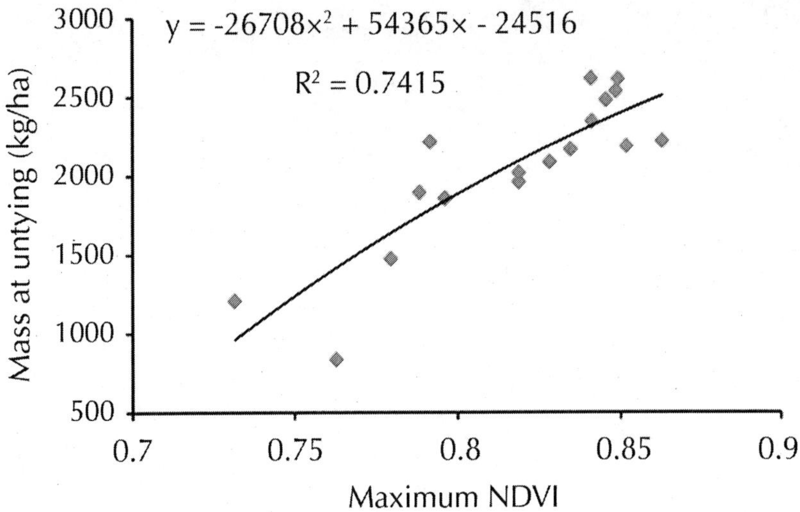
y = -26708x² + 54365x - 24516
R² = 0.7415
Mass at untying (kg/ha)
3000
2500
2000
1500
1000
500
0.7
0.75
0.8
0.85
0.9
Maximum NDVI

(a)

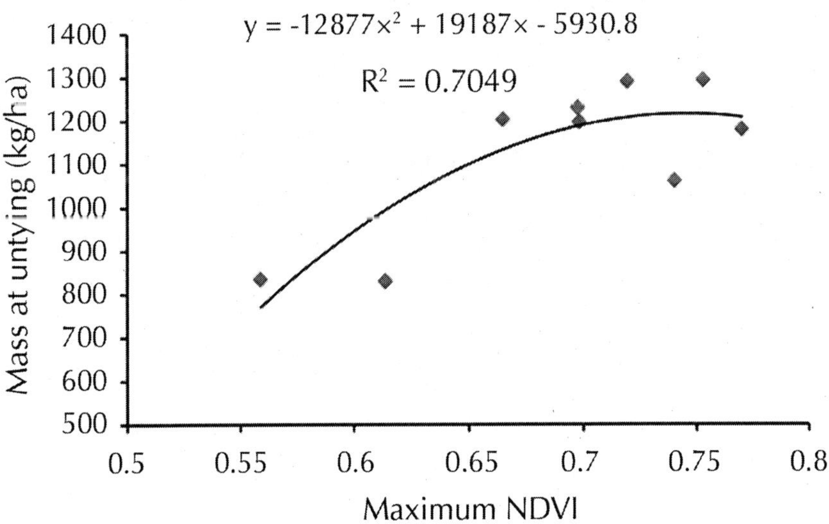
y = -12877x² + 19187x - 5930.8
R² = 0.7049
Mass at untying (kg/ha)
1400
1300
1200
1100
1000
900
800
700
600
500
0.5
0.55
0.6
0.65
0.7
0.75
0.8
Maximum NDVI

(b)

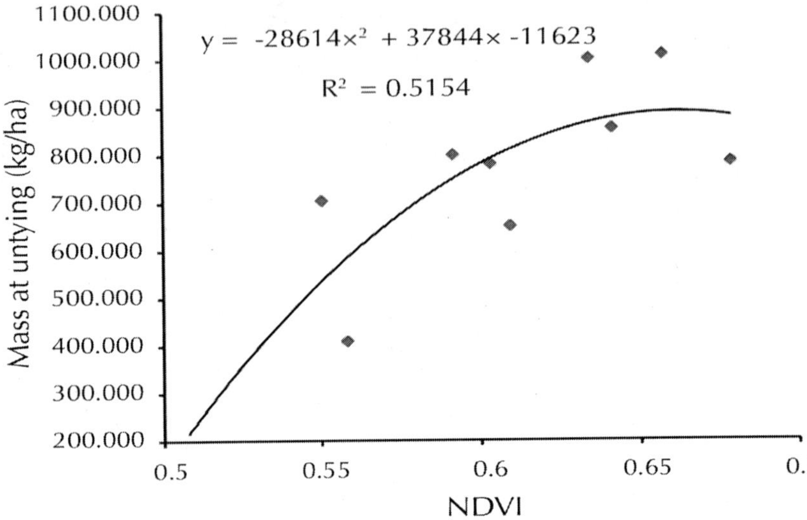


(c)

**Figure 7:** The relationship between tobacco mass at untying for the (a) September-October, (b) November, and (c) December planted crops and maximum NDVI.

# DISCUSSIONS

The October crop was selected for in-season dry mass-NDVI analysis because of the clear cloudless conditions during the times of data collection. Thin cloud coverage, according to Nuarsa et al. [29], can lead to inconsistencies in the reflectance values, which will affect the NDVI, and, therefore, the selection of cloud-free images is one of the most important steps in the data analysis.

The NDVI-yield relationship increases with age up to 10–12-week after planting. As Nageswara-Rao et al. [30] explained it, the changes in spectral response of a crop are a function of phenological stages of the crop. The 10–12 weeks period, with the highest correlation, could be an indication of the most suitable phenological stage to collect satellite data for yield forecasting.

Chlorophyll degradation related leaf ripening occurring during the ripening stage causes an increase in the red spectral reflectance which is normally absorbed by chlorophyll [31]. On the contrary, the NIR spectral reflectance is decreased due to a change in leaf internal structure [31] resulting in the fall of NDVI [32]. The fall in the in-season dry-mass-NDVI relationship after 14 weeks of age is related to the decrease in canopy reflectance spectra decrease at crop maturity stage that is brought about by reaping [33], while the final yield, in the data analysis, remains unchanged.

The decrease in tobacco mass at untying with later planting at all variety × fertiliser treatments was long since established [10]. Apparently the maximum NDVI in this experiment also followed the same trend, indicating a positive relationship between the two.

The similarity in the coefficients of determination between mass at untying and NDVI for the September and October planted crops could be an indication of the need to combine the three, when assessing area and yield using remote sensing, while the November and the December crops could each be assessed separately. The high coefficients of determination for all the three varieties and fertilizer levels could also be an indication of the possibility to disregard the variety and fertiliser differences in the processes of developing yield forecasting models.

The coefficients of determination between mass at untying and NDVI for the September-October (0.741) and the November (0.704) planted crops were higher than the 0.65 reported by Povkh et al. [34] but lower than the $r^2$ = 0.90–0.98 that Jiang et al. [35] found between wheat grain yield and NDVI. The established coefficients in this experiment were, however, high enough for tobacco yield to be estimated using Cropscan calculated NDVI. The yield models derived were quadratic, similar to the findings of Jiang et al. [35] in wheat. The high value of $R^2$ indicated that the relationship between tobacco yield and the NDVI was consistent [29]. The December crop coefficient of determination was, however, low (0.515) meaning that yield estimation for this crop would not be accurately made using the model.

As the channels of the sensor used in the experiment is LANDSAT Thematic Mapper compatible [36], the models derived can be applicable in tobacco yield estimation using operation remote sensing data from the satellite.

However, more work is needed to establish the relationship between the Cropscan reflectance and those for selected Satellite platforms like Modis, Landsat 5 and Landsat TM which have been used for the same purpose in other crops [37].

## CONCLUSIONS

The NDVI is positively related to in-season dry mass and, can be used to assess crop health, tobacco response to fertilizer, and accurate monitoring of crop development patterns for yield forecasting purposes. For yield forecasting purposes, the September and the October crops could be estimated together, while the November and the December crops each could also be estimated, separately. There was a strong positive correlation between NDVI and flue-cured tobacco yield at all fertilizer levels and for all the tested varieties, and, hence, for yield forecasting purposes, one may not separate these factors. There is, however, a need to establish the relationship between Cropscan multispectral radiometer 5 data and various satellite platforms before this information can be applied in satellite remote sensing.

## ACKNOWLEDGMENTS

The authors are grateful to the Tobacco Research Board/Kutsaga Research Station for funding this series of experiments on developing-flue cured tobacco crop area and yield forecasting models using remote sensing and agronomic techniques.

## REFERENCES

1. C. A. Reynolds, M. Yitayew, D. C. Slack, C. F. Hutchinson, A. Huetes, and M. S. Petersen, "Estimating crop yields and production by integrating the FAO Crop Specific Water Balance model with real-time satellite data and ground-based ancillary data," International Journal of Remote Sensing, vol. 21, no. 18, pp. 3487–3508, 2000.

2. S. Landau, R. A. C. Mitchell, V. Barnett, J. J. Colls, J. Craigon, and R. W. Payne, "A parsimonious, multiple-regression model of wheat yield response to environment,"Agricultural and Forest Meteorology, vol. 101, no. 2-3, pp. 151–166, 2000
3. T. R. Wheeler, P. Q. Craufurd, R. H. Ellis, J. R. Porter, and P. V. Vara Prasad, "Temperature variability and the yield of annual crops," Agriculture, Ecosystems and Environment, vol. 82, no. 1–3, pp. 159–167, 2000
4. SADC, Selected Technical Papers: Methodology of Food Crop Forecasting in SAD, SADC Secretariat, Gaborone, Botswana, 2009.
5. P. C. Doraiswamy, B. Akhmedovb, L. Beardc, A. Sterna, and R. Mueller, "Operational prediction of crop yields using MODIS data and products," in Proceedings of the International Achives of Photogrametry, Remote Sensing and Spatial Information Sciences, vol. 38, pp. 45–50, 2011.
6. F. Tao, M. Yokozawa, Z. Zhang, Y. Xu, and Y. Hayashi, "Remote sensing of crop production in China by production efficiency models: models comparisons, estimates and uncertainties," Ecological Modelling, vol. 183, no. 4, pp. 385–396, 2005
7. P. C. Doraiswamy, J. L. Hatfield, T. J. Jackson, B. Akhmedov, J. Prueger, and A. Stern, "Crop condition and yield simulations using Landsat and MODIS," Remote Sensing of Environment, vol. 92, no. 4, pp. 548–559, 2004.
8. F. Rembold, C. Atzberger, I. Savin, and O. Rojas, "Using low resolution satellite imagery for Yield prediction. and yield anomaly detection," Remote Sensing, vol. 5, no. 4, pp. 1704–1733, 2013
9. A. D. Baez-Gonzalez, J. R. Kiniry, S. J. Maas et al., "Large-area maize yield forecasting using leaf area index based yield model," Agronomy Journal, vol. 97, no. 2, pp. 418–425, 2005.
10. Tobacco Research Board (TRB), Flue Cured Tobacco Recommendations, TRB, Harare, Zimbabwe, 2012.
11. C. J. Tucker, "Red and photographic infrared linear combinations for monitoring vegetation," Remote Sensing of Environment, vol. 8, no. 2, pp. 127–150, 1979.

12. J. Verrelst, B. Koetz, M. Kneubühler, and M. Schaepman, "Directional sensitivity analysis of vegetation indices from multiangular CHRIS/PROBA data," in Proceedings of the ISPRS Commission VII Mid-Term Symposium, N. Kerle and A. Skidmore, Eds., Enschede, The Netherlands.
13. D. Gross, Monitoring Agricultural Biomass Using NDVI Time Series, Food and Agriculture Organization of the United Nations (FAO), Rome, Italy, 2005.
14. A. Viña, A. A. Gitelson, A. L. Nguy-Robertson, and Y. Peng, "Comparison of different vegetation indices for the remote assessment of green leaf area index of crops," Remote Sensing of Environment, vol. 115, no. 12, pp. 3468–3478, 2011.
15. S. C. Liew, "Principles of Remote Sensing," Centre for Remote Imaging, Sensing and Processing, National University of Singapore, Singapore, 2001,http://www.crisp.nus.edu.sg/~research/tutorial/rsmain.htm.
16. J. T. Woolley, "Reflectance and Transmittance of Light by Leaves," Plant Physiology, vol. 47, pp. 656–662, 1970.
17. A. K. Prasad, L. Chai, R. P. Singh, and M. Kafatos, "Crop yield estimation model for Iowa using remote sensing and surface parameters," International Journal of Applied Earth Observation and Geoinformation, vol. 8, no. 1, pp. 26–33, 2006.
18. J. U. H. Eitel, R. F. Keefe, D. S. Long, A. S. Davis, and L. A. Vierling, "Active ground optical remote sensing for improved monitoring of seedling stress in nurseries," Sensors, vol. 10, no. 4, pp. 2843–2850, 2010
19. X. Yin, A. McClure, and D. Tyler, "Relationships of plant height and canopy NDVI with nitrogen nutrition and yields of corn," in Proceedings of the 19th World Congress of Soil Science, Soil Solutions for a Changing World, Brisbane, Australia, August 2010.
20. R. D. Jackson, P. N. Slater, and P. J. Pinter Jr., "Discrimination of growth and water stress in wheat by various vegetation indices through clear and turbid atmospheres," Remote Sensing of Environment, vol. 13, no. 3, pp. 187–208, 1983.
21. C. S. Muthy, S. Jonna, P. V. Raju, S. Thurivengadachari, and K. A. Hakeem, "Crop Yield Prediction in Command Area using Satellite Data," GISdevelopment.net, AARS, ACRS 1994, Poster Session.

22. T. Engel, G. Hoogenboom, J. W. Jones, and P. W. Wilkens, "AEGIS/WIN: a computer program for the application of crop simulation models across geographic areas," Agronomy Journal, vol. 89, no. 6, pp. 919–928, 1997
23. Tobacco Industries Marketing Board (T.I.M.B), Annual Report and Accounts For the Year Ended 30 June, 2011, T.I.M.B, Harare, Zimbabwe, 2012.
24. C. Atzberger, "Advances in remote sensing of agriculture: context description, existing operational monitoring systems and major information needs," Remote Sensing, vol. 5, pp. 949–981, 2012.
25. C. Ferencz, P. Bognár, J. Lichtenberger et al., "Crop yield estimation by satellite remote sensing," International Journal of Remote Sensing, vol. 25, no. 20, pp. 4113–4149, 2004
26. S. A. Mohammad, "Spectral indices and agronomic variables relationship of cotton (Gossypium sps.) Under sowing dates and nitrogen levels," Asian Conference for Remote Sensing, 2008,http://www.gisdevelopment.net/application/agriculture/yield/mi08_305.htm.
27. K. Dalsted and L. Queen, "Interpreting remote sensing data," The Site-Specific Management Guideline SSMG-26.
28. H. Zhang, H. Chen, and G. Zhou, "The model of wheat yield forecast based on modis-ndvi: a case study of xinxiang," in Proceedings of the ISPRS Annals of the Photogrammetry, Remote Sensing and Spatial Information Sciences Congress, Melbourne, Australia, August 2012.
29. I. W. Nuarsa, F. Nishio, and H. Chiharu, "Rice yield estimation using landsat ETM+ data and field observation," Journal of Agricultural Science, vol. 4, no. 3, pp. 45–56, 2012.
30. R. C. Nageswara-Rao, J. H. Williams, M. V. K. Sivakumar, and K. D. R. Wadia, "Effect of water deficit at different growth phases of groundnut. II. Response to drought during pre-flowering phas," Agronomy Journal, vol. 80, pp. 431–438, 1988.
31. W. Gunnula, M. Kosittrakun, T. L. Righetti, P. Weerathaworn, and M. Prabpan, "Normalized difference vegetation index relationships with rainfall patterns and yield in small plantings of rain-fed sugarcane," Australian Journal of Crop Science, vol. 5, no. 13, pp. 1852–1857, 2011

32. D. M. Gates, H. J. Keegan, J. C. Schleter, and V. R. Weidner, "Spectral properties of plants,"Applied Optics, vol. 4, no. 1, pp. 11–20, 1965.

33. H. Qiao, W. Mei, Y. Yang, W. Yong, J. Zhang, and Y. Hua, "Study on relationship between tobacco canopy spectra and LAI," IFIP Advances in Information and Communication Technology, vol. 345, no. 2, pp. 650–657, 2011

34. V. Povkh, G. L. Shljakhova, Garbuzov, and E. Vorobeychik, "Operational monitoring of the agricultural production based on the observational MODIS data as a support for improving regional planning," in Proceedings Of The International Symposium On Remote Sensing Of Environment, South Regional Information and Analytical Center (SRIA-Center), Rostov-on-Don, RUSSIA, 2005.

35. D. Jiang, N.-B. Wang, X.-H. Yang, and J.-H. Wang, "Study on the interaction between NDVI profile and the growing status of crops," Chinese Geographical Science, vol. 13, no. 1, pp. 62–65, 2003

36. I. C. T. International, Multispectral Radiometers, ICT International, 2003.

37. C. Yang, J. H. Everitt, and J. M. Bradford, "Using high resolution QuickBird satellite imagery for cotton yield estimation," in Proceedings of the ASAE Annual International Meeting, pp. 893–904, August 2004.

Chapter 7

# Evaluation of Phosphorus Fertilizer Rates for Maize and Sources for Cowpea on Differe Soil Types in Southwestern Nigeria

G. O. Kolawole, A. O. Olayiwola, O. Ige, G. O. Oyediran, and B. A. Lawal

Department of Agronomy, Ladoke Akintola University of Technology (LAUTECH), PMB 4000, Ogbomoso, Oyo State, Nigeria

## ABSTRACT

Flexible phosphorus (P) fertilizer rate recommendation could be based on variations in soil characteristics that affect yield responses. Experiments were conducted in the Department of Agronomy, LAUTECH, on the effects of P rates on maize and P sources on cowpea in four soil types. On average, soil types and P rates influenced maize height and grain yield. Iwo and Egbeda soils supported taller plants than Itagunmodi soil. Phosphorus fertilization enhanced height and grain yield compared with no P. To optimize maize grain yield for

Itagunmodi and Egbeda soils, application of 15 kg $P_2O_5$ $ha^{-1}$ was sufficient while for Majeroku and Iwo, it was 30 and 75 kg $P_2O_5$ $ha^{-1}$, respectively. Cowpea grain yield and P uptake were significantly affected by soil types and P sources. Iwo and Egbeda soils supported higher grain weights and P uptake than Itagunmodi and Majeroku soils. Triple super phosphate (TSP) and no P supported higher grain weights and P uptake than rock phosphate (RP) and single super phosphate (SSP).

# INTRODUCTION

The subject of soil parameter variation within fields is becoming an important topic in the agronomic community. Beckett and Webster (1971) presented a review of lateral variability of soil properties. They found that up to onehalf of the variance within a field might be present in as little as one square meter of land. Recognition of this variability has prompted many researchers to consider managing this variability. Carr et al. (1991) suggested, "Farming soils, not fields". Their study was initiated to measure crop yield differences between contrasting soils within a field and to compare the economics of varying nutrient application by contrasting soils with the traditional practice of field-average application. Returns to farming soils were generally greater than when farming fields. However, the researchers noted that it was essential to establish appropriate crop yield goals, conduct accurate soil tests and utilize reliable fertilizer recommendations to generate greater returns when managing nutrients. Smallholder farmers effectively deal with soil variation by location-specific field management based on crop performance and crop responses they observed in their fields over past years. Research that aims to improve soil fertility management and productivity of small-scale farmers has to reckon with soil variation and has to come up with flexible recommendations rather than blanket recommendations. Blanket recommendations may raise the 'average' productivity in an area, but yield negative responses on parts of the fields and farms and therefore, discredit extension messages. Flexible recommendation could be based on variations in soil characteristics that affect productivity and yield responses.

Phosphorus is a major limiting factor for crop production on many tropical and sub-tropical soils (Norman et al., 1995) as a result of high

P fixation and/or nutrient mining in traditional land-use systems, due to poor accessibility and high cost of soluble P fertilizers for a large population of disadvantaged farmers. Phosphorus deficiency is so acute that plant growth ceases as soon as the phosphorus deficiency is so acute that plant growth ceases as soon as the phosphorus stored in the seed is exhausted in some soils of the savanna zone of Western Africa (Mokwunye et al., 1986). Consequently, they require the addition of P fertilizers for producing even moderate yields. Numerous studies have shown that P fertilizers can significantly increase crop yields (Batiano et al., 1995; Kolawole et al., 2000). Crop response to P fertilizer application however depend on many factors, such as, soil characteristics, crop grown, climate, tillage systems, interactions with other nutrients, crop management and fertilizer management. It is therefore necessary to take these factors into consideration before embarking on P fertilization programme to improve fertilizer use efficiency and economic returns. In many parts of the tropics, large spatial variability in soil characteristics occurs within short distances. Under such situation, it becomes inevitable that fertilizer recommendations should be site specific.

**Table 1:** Physico-chemical properties of four soil types in southwestern Nigeria

| Properties | Itagunmodi | **Egbeda** | **Majeroku** | **Iwo** |
|---|---|---|---|---|
| Sand (g $kg^{-1}$) | 590 | 690 | 810 | 690 |
| Silt (g $kg^{-1}$) | 150 | 190 | 90 | 130 |
| Clay (g $kg^{-1}$) | 260 | 120 | 100 | 180 |
| Soil pH-$H_20$ | 4.5 | 5.3 | 4.9 | 5.3 |
| Ex.Ca (cmol $kg^{-1}$) | 2.05 | 2.12 | 0.96 | 5.33 |
| Ex. Mg (cmol $kg^{-1}$) | 0.66 | 0.49 | 0.36 | 0.53 |
| Ex. K (cmol $kg^{-1}$) | 0.14 | 0.12 | 0.16 | 0.29 |
| Ex. Na (cmol $kg^{-1}$) | 0.31 | 0.31 | 0.37 | 0.36 |
| Ex. Fe (ppm) | 149.78 | 156.46 | 138.67 | 184.89 |
| Available P (µg $g^{-1}$) | 6.84 | 41.88 | 9.64 | 32.18 |
| Total N (g $kg^{-1}$) | 1.11 | 0.52 | 0.37 | 1.16 |

| Organic C (g $kg^{-1}$) | 10.84 | 5.39 | 3.41 | 12.46 |
|---|---|---|---|---|

Fertilizer recommendations in Nigeria have generally been based on research results from a limited number of sites (Orkwor and Asadu, 1998). Hence, such 'blan ket recommendations' can be misleading when transferred to another ecology.

Variable P fertilizer rates and sources management can improve both fertilizer use efficiency and economic returns. The objective of the present study is to provide an insight into the appropriate P fertilizer rates for maize and the appropriate P fertilizer sources for cowpea in the dominant soil types of the moist savanna of southwestern Nigeria.

# MATERIALS AND METHODS

The experiments were conducted in the Department of Agronomy, Ladoke Akintola University of Technology (LAUTECH) in Ogbomoso (Longitude 4° 10′ E, Latitude 8° 10′ N and altitude 213 m asl), Oyo State, Nigeria, during May - August 2006 and February - April 2007, respectively.

## Experiment 1: Effects of Soil Types and P Fertilizer Rates on the Performance of Maize

Surface soil (0 – 15 cm depth) of four soil types (Itagunmodi (Rhodic Paleutult), Iwo (Oxic Haplustalf), Egbeda (Oxic Paleustult) and Majeroku (Abruptic Tropaqualf) series were collected from four study sites in Oyo and Osun States, Nigeria. The soils were airdried, sieved through a 2 mm sieve and subsamples were taken for laboratory analysis. The physico-chemical characteristics of the soils are presented in Table 1. Smyth and Montgomery (1962) described in details the characteristics of these soils.

The four soil types formed the main plot treatments and six P rates: 0 (control), 15, 30, 60, 75 and 90 kg $P_2O_5$ ha $^{-1}$ applied as single super phosphate (SSP) were the sub treatments. The treatments were replicated four times and arranged in completely randomised design. Fifteen kilogram soil each was weighed into 24 pots for each of the four soil types. This makes a total of 96 pots. The soil was watered to field capacity and left to equilibrate for 48 h before planting. Three seeds each of maize variety DMR-L-SR were sown in the middle of the pots and the seedlings were later thinned to one per pot at two weeks after emergence. Nitrogen fertilizer in form of urea and potassium fertilizer in the form of muriate of potash were applied to all the pots in equal amounts (120 kg N ha $^{-1}$ that is, 2 g Urea 15 kg soil -1 ) and (60 kg $K_2O$ ha $^{-1}$ that is, 0.8 g MOP 15 kg soil $^{-1}$ ) as basal dressing. The plants were watered regularly as necessary. Weeds were hand pulled as they emerged and left in the pots to decompose. Maize height was measured fortnightly starting from 2 weeks after planting (WAP) up to 10 WAP. At physiological maturity, dry grain yield was determined at 15% moisture content using moisture tester.

## Experiment 2: Effects of Soil Types and P Fertilizer Sources on Performance of Cowpea

The four soil types used in experiment 1 above formed the main plot treatments and four P sources: no P (control), phosphate rock (RP) (90 kg $P_2O_5$ ha $^{-1}$), triple super phosphate (TSP) (30 kg $P_2O_5$ ha $^{-1}$) and single super phosphate (SSP) (30 kg $P_2O_5$ ha $^{-1}$) were the sub treatments. The treatments were replicated three times and arranged in completely randomised design. Higher rates of RP compared to SSP and TSP was used because Zapata (1986) reported that the relative P effectiveness of rock P is about 3 - 5

Times less than that of artificial P-fertilizers on annual crops. Four kilogram soil each was weighed into 12 pots for each of the four soil types. This makes a total of 48 pots used for the experiment. The soil was watered to field capacity and left to equilibrate for 48 h before planting. Five seeds each of cowpea var. IT 90K – 277 – 2 were sown in each pot and the seedlings thinned to one per pot at two weeks after emergence. The fertilizers were applied in a ring form about 5 cm away from the cowpea plant at the time of thinning. The plants were watered regularly as necessary. Weeds were hand pulled as they emerged and left in the pots to decompose. Insect attack was controlled by mixing one ml of karate ® 2.5 E.C. (a.i. 25 g lambda-cyhalothrin per liter) insecticide into 500 ml of water to spray the plants. This was done four times before harvesting of cowpea. At physiological maturity, the number of pods was counted, the pods were harvested and the grains removed. The shoot was cut at ground level with a sharp knife. Litter was collected as part of the shoot biomass. The leaves and petioles were separated from the stem. All plant parts were placed in separate paper bags and oven dried at 65°C for 72 h to determine their dry weights. The grains were ground to pass through a 2 mm mesh sized sieve and analysed for P concentrations. For determination of P, samples were wet–digested with a mixture of $HClO_{4-}$ $HNO_3$. Phosphorus was measured calorimetrically by auto-analyzer (IITA, 1982).

All data were subjected to analysis of variance (ANOVA). Where F-values were significant, the treatment means were separated with least significant difference (LSD) test at the 5% probability level. Response curves for maize grain weight to P rates for the soil types was developed using regression analysis.

# RESULTS

## Experiment 1

### *Maize Plant Height*

At the early stage (2 weeks after planting) (WAP) of growth, the soil types had no significant effects on height of maize plants. However, at 4 and 8 WAP, Iwo and Egbeda soils supported taller plants than It a gunmodi soil and also at 6 and 10 WAP, maize plants grown on Itagunmodi soil were significantly shorter than those grown on the other three soil types (Table 2). Throughout the sampling period, application of P resulted in taller plants than the no P situation. For most of the periods, the highest P rate (90 kg $P_2O_5$ ha $^{-1}$) supported the tallest maize plants. Interaction effect between soil types and P rates on plant height was not significant.

### *Maize Dry Grain Yield*

Soil types did not significantly affect maize dry grain yields, but application of P fertilizer resulted in higher grain yield than the no P situation. Apart from the control, the other P rates had similar effects on grain yield. However, interaction effects between soil types and P rates were significant (Figure 1). For example, for Itagunmodi and Egbeda soil types, application of 15 kg $P_2O_5$ ha $^{-1}$ appeared sufficient in optimising grain yield while application of 30 kg $P_2O_5$ ha $^{-1}$ seemed sufficient for Majeroku soil type. Whereas, for Iwo soil type, application of 75 kg $P_2O_5$ ha $^{-1}$ appeared adequate in influencing optimum maize grain yield under the condition of this trial.

## Experiment 2

### ***Cowpea Dry Grain Weight***

On the average, soil types and P fertilizer sources had significant effects on dry grain weight of cowpea (Table 3). Iwo (26.4 g plant $^{-1}$) and Egbeda (22.3 g plant $^{-1}$) soil types supported significantly higher grain weights than Itagunmodi (13.6 g plant $^{-1}$) and Majeroku (12.3 g plant $^{-1}$) soil types. TSP (24.4 g plant $^{-1}$) and no P treatment (21.2 g plant $^{-1}$) supported significantly higher grain weights than RP (15.6 g plant $^{-1}$) and SSP (13.3 g plant $^{-1}$). Soil types and P sources interaction effects on grain weight was significant. For example, P application had negative effects on grain weights under Iwo and Majeroku soil types whereas, under Itagunmodi and Egbeda soil types, grain weights were higher with application of RP compared with the no P situation and application of TSP profoundly improved grain weights compared with other P sources on these soils. Application of SSP under Iwo soil type resulted in higher grain weight than those of RP and TSP. For Itagunmodi and Egbeda soil types, application of SSP influenced lower grain weights than those of RP and TSP.

### ***Shoot Dry Weight***

Similar to the results obtained for grain weights and soil types, P sources on the average, significantly affected shoot dry weight of cowpea. Iwo soil type supported significantly higher shoot biomass than the other soil types and TSP was superior to the other P sources in their effects on shoot dry weights (Table 3). Soil types and P sources interaction effects was significant. Application of TSP influenced higher shoot biomass production in the soils except for Egbeda soil type where RP and TSP had similar effects on shoot weights.

## ***Phosphorus Concentration and Uptake in Cowpea Grain***

On the average, soil types and P fertilizer sources had significant effects on cowpea grain P concentrations and uptake. Iwo soil type influenced significantly higher P concentration than Egbeda and Itagunmodi soil types. Egbeda soil type influenced significantly higher P concentration than Majeroku soil type. Application of TSP supported significantly the highest P concentration compared with the other P sources while the no P situation and RP treatments had similar P concentrations. The no P situation however supported significantly higher P concentration than application of SSP (Table 4). Soil types and P sources interaction effects on P concentrations of cowpea grain was not significant.

**Table 2:** Effects of soil types and P fertilizer application rates on maize plant height (cm) at Ogbomoso, southwestern Nigeria

| Soil types | Plant height (cm) | | | | | | |
|---|---|---|---|---|---|---|---|
| | 0 | 15 kg P ha $^{-1}$ | 30 kg P ha $^{-1}$ | 60 kg P ha $^{-1}$ | 75 kg P ha $^{-1}$ | 90 kg P ha $^{-1}$ | Soil type means |
| 2 Weeks after planting (WAP) | | | | | | | |
| Itagunmodi | 6.8 | 7.1 | 9.1 | 9.9 | 9.0 | 10.1 | 8.7 |
| Egbeda | 6.5 | 8.5 | 8.1 | 8.4 | 8.4 | 9.8 | 8.3 |
| Majeroku | 6.6 | 6.4 | 8.5 | 8.0 | 7.5 | 9.5 | 7.8 |
| Iwo | 6.1 | 7.6 | 10.5 | 8.3 | 9.6 | 9.1 | 8.5 |
| P rate means | 6.5 | 7.4 | 9.1 | 8.6 | 8.6 | 9.6 | |
| **4 WAP** | | | | | | | |
| Itagunmodi | 7.9 | 15.9 | 19.6 | 21.8 | 16.9 | 24.3 | 17.7 |
| Egbeda | 14.4 | 19.9 | 21.5 | 24.1 | 23.6 | 30.5 | 22.3 |
| Majeroku | 14.3 | 19.1 | 23.5 | 18.4 | 21.9 | 25.3 | 20.4 |
| Iwo | 18.6 | 22.5 | 23.9 | 25.6 | 25.3 | 26.4 | 23.7 |
| P rate means | 13.8 | 19.3 | 22.1 | 22.5 | 21.9 | 26.6 | |
| **6 WAP** | | | | | | | |
| Itagunmodi | 17.3 | 31.4 | 41.5 | 46.8 | 35.0 | 57.8 | 38.3 |
| Egbeda | 28.6 | 37.5 | 46.8 | 48.6 | 52.3 | 62.9 | 46.1 |
| Majeroku | 29.6 | 40.1 | 56.5 | 47.5 | 47.5 | 64.5 | 47.6 |
| Iwo | 36.0 | 42.3 | 50.5 | 58.5 | 59.8 | 66.5 | 52.3 |
| P rate means | 27.9 | 37.8 | 48.8 | 50.3 | 48.6 | 62.9 | |
| **8 WAP** | | | | | | | |
| Itagunmodi | 33.0 | 83.5 | 105.8 | 154.5 | 94.8 | 141.5 | 102.2 |
| Egbeda | 59.5 | 91.5 | 119.3 | 123.8 | 153.5 | 147.0 | 115.8 |
| Majeroku | 70.5 | 93.3 | 125.8 | 115.0 | 120.3 | 146.0 | 111.8 |
| Iwo | 97.0 | 102.3 | 133.8 | 147.3 | 146.8 | 162.8 | 131.6 |
| P rate means | 65.0 | 92.6 | 121.1 | 135.1 | 128.8 | 149.3 | |
| **10 WAP** | | | | | | | |
| Itagunmodi | 75.8 | 132.8 | 154.0 | 173.3 | 118.5 | 161.3 | 135.9 |
| Egbeda | 127.0 | 143.8 | 183.8 | 160.3 | 184.5 | 184.0 | 163.9 |
| Majeroku | 164.0 | 166.3 | 183.8 | 168.8 | 170.5 | 169.3 | 168.8 |
| Iwo | 151.8 | 169.8 | 173.0 | 169.3 | 178.8 | 189.5 | 167.0 |
| P rate means | 129.6 | 143.1 | 173.6 | 167.9 | 163.1 | 176.0 | |

LSD $_{(0.05)}$: Soil types: 1.06; 2.86; 6.71; 17.57; 14.86 for 2, 4, 6, 8 and 10 WAP. LSD$_{+ (0.05)}$: P rates: 1.29; 3.50; 8.22; 21.52; 18.20 for 2, 4, 6, 8 and 10 WAP.

**Table 3:** Effects of soil types and P fertilizer sources on cowpea dry grain and shoot weights at Ogbomoso, southwestern Nigeria 2007

| Soil types | Dry grain weights (g plants$^{-1}$) | | | | | Dry shoot weights (g plants$^{-1}$) | | | | |
|---|---|---|---|---|---|---|---|---|---|---|
| | 0P | RP | SSP | TSP | Soil type means | 0P | RP | SSP | TSP | Soil type means |
| Iwo | 46.0 | 18.7 | 24.3 | 16.3 | 26.4 | 53.7 | 58.8 | 21.5 | 145.0 | 69.7 |
| Itagunmodi | 9.1 | 14.7 | 7.0 | 23.5 | 13.6 | 11.8 | 5.4 | 43.1 | 58.9 | 29.8 |
| Egbeda | 11.6 | 20.7 | 11.4 | 45.6 | 22.3 | 17.5 | 48.0 | 7.2 | 46.2 | 29.7 |
| Majeroku | 17.9 | 8.5 | 10.6 | 12.3 | 12.3 | 30.5 | 16.7 | 9.4 | 33.8 | 22.6 |
| P source means | 21.2 | 15.7 | 13.3 | 24.4 | | 28.4 | 32.2 | 20.3 | 71.0 | |

Grain weight: LSD $_{(0.05)}$ soil types (S): 8.7; LSD (0.05) P sources (P): 5.2; LSD $_{(0.05)}$ SxP: 10.4. Shoot weight: 19.0, 20.0 and 40.0.

Iwo and Egbeda soil types influenced higher cowpea grain P uptake than Itagunmodi and Majeroku soil types. Application of TSP and the no P situation had higher P uptake than application of RP and SSP but RP was superior to SSP in their effects on P uptake. Soil types and P sources interaction effects on P uptake of cowpea grain was significant. For example, on Iwo and Majeroku soil types, the no P situation had highest P uptake values while for Egbeda and Itagunmodi soil types, TSP treatment had the highest P uptake values. In all the soil types (except Iwo), P uptake values were the least with application of SSP (Table 4).

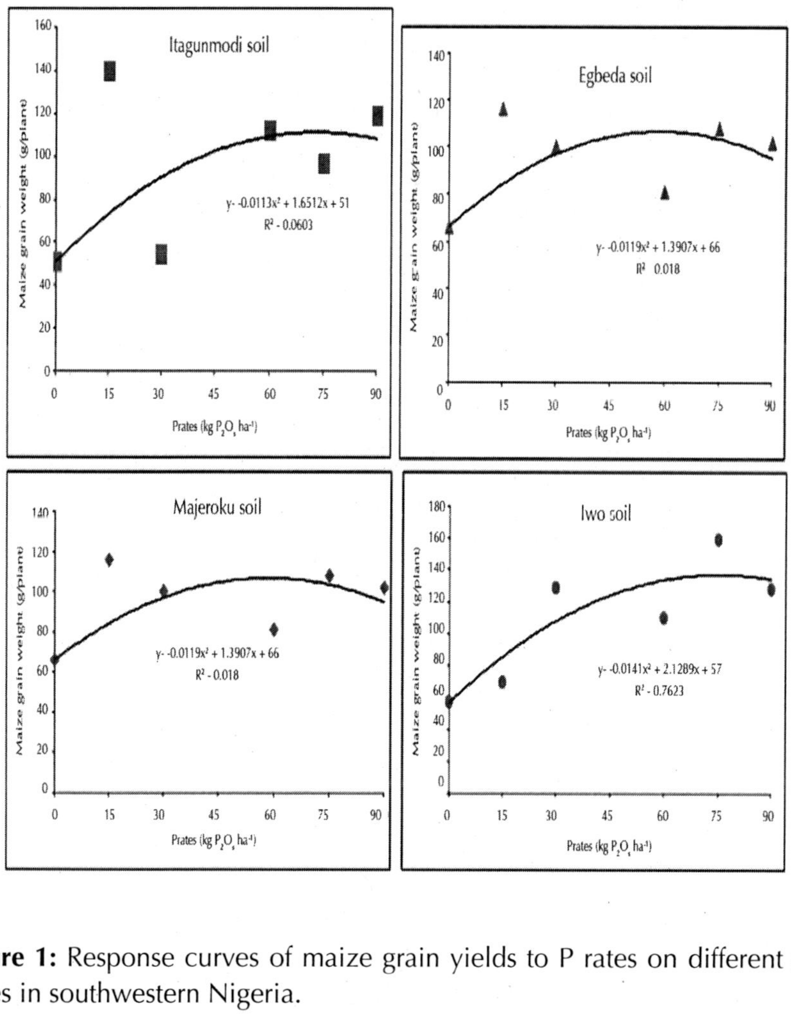


**Figure 1:** Response curves of maize grain yields to P rates on different soil types in southwestern Nigeria.

**Table 4:** Effects of soil types and P fertilizer sources on cowpea grain P concentrations and uptake at Ogbomoso, south-western Nigeria 2007

| Soil types | P contents (%) | | | | | P uptake (mg plant$^{-1}$) | | | | |
|---|---|---|---|---|---|---|---|---|---|---|
| | 0P | RP | **SSP** | **TSP** | **Soil type means** | **0P** | **RP** | **SSP** | **TSP** | **Soil type means** |
| Iwo | 0.308 | 0.234 | 0.236 | 0.379 | 0.289 | 1.46 | 0.42 | 0.50 | 0.54 | 0.73 |
| Itagunmodi | 0.224 | 0.260 | 0.155 | 0.337 | 0.244 | 0.20 | 0.41 | 0.11 | 0.52 | 0.31 |
| Egbeda | 0.249 | 0.277 | 0.178 | 0.316 | 0.255 | 0.17 | 0.48 | 0.12 | 1.10 | 0.47 |
| Majeroku | 0.274 | 0.196 | 0.192 | 0.259 | 0.230 | 0.37 | 0.17 | 0.13 | 0.33 | 0.25 |
| P source means | 0.264 | 0.242 | 0.323 | 0.190 | | 0.55 | 0.37 | 0.21 | 0.62 | |

P concentrations: $LSD_{(0.05)}$ soil types (S): 0.016; $LSD_{(0.05)}$ P sources (P): 0.057; $LSD_{(0.05)}$ SxP: ns. P uptake: 0.34, 0.14, 0.43.

# DISCUSSION

The available data indicated that the soil types used in this study varied in soil available P. Whereas soil available P was low in Itagunmodi and Majeroku soils, Egbeda and Iwo soils have high soil available P. This probably influenced the response of maize to varied P rates in the different soils. Although the trend fluctuations of maize grain weights with changes in P rates were inconsistent for the soil types (except for Iwo soil) (Figure 1), nonetheless, there were indications of the P rates at which highest grain weights were observed for the different soil types under the conditions of this experiment. For example, to optimize maize grain yield for Itagunmodi and Egbeda soil types' application of 15 kg $P_2O_5$ ha $^{-1}$ appeared sufficient while for Majeroku and Iwo, it was 30 and 75 kg $P_2O_5$ ha $^{-1}$ , respectively. This is an indication that different amounts of fertilizer inputs are needed to achieve similar yields on different soil types, indicating that response to applied inputs or their agronomic use efficiency is likely to be affected by soil characteristics that affect productivity and yield responses. McKenzie et al. (2003) reported that soil types affected P fertilizer response by different crops in Canada. Sanchez and Uehara (1980) stated that the soil solution P concentration required for maximum plant growth differs with soil properties related to the diffusion of P to plant roots. In the present study, Iwo soil which had high soil available P also require the highest P rate to obtain optimum maize grain. This may be due to its highest Ca and Fe contents which may make applied P unavailable for plant use through fixation. Availability of soil P is a complex phenomenon which is affected by both soil and plant properties (Barber, 1995). Even when P fertilizers are used, it may become adsorbed in various P compounds of low solubility, hence it is unavailable to the crop (Holford, 1997). The poor performances of both maize and cowpea crops in Itagunmodi soil could be due to its low available P content and high clay content which encourages caking, leading to poor root development. Javid (1999) reported that clay contents showed best correlation with P adsorbing capacities of soils. Conversely, Iwo and Egbeda soils which influenced good crop performance had lower clay contents and contained high available P contents available P contents. It is quite interesting to note that for cowpea, fertilizer P sources that promote good performance differ for the different soil types. For example, P application had negative effects on grain weights under Iwo and Majeroku soil types whereas, under Itagunmodi and Egbeda soil types, grain weights were higher with application of RP compared with the no P situation and application of TSP profoundly improved grain weights compared with other P sources on these soils. Application of SSP under Iwo soil type resulted in higher grain weight than those of RP and TSP. For Itagunmodi and Egbeda soil types, application of SSP influenced lower grain weights than those of RP and TSP. It has been reported that phosphorus availability was differentially influenced by different P sources and different soils (Torres-Dorante et al., 2006). Van Ray and Van Diest (1979) also noted that for three P sources, super phosphate, calcium aluminium phosphate and hyperphosphate, a relationship between soil acidity and P uptake was found.

ptake was found. From the results of this study, it can be deduced that blanket P fertilizer recommendations can be misleading and that variable P fertilizer rates and sources management can improve both fertilizer use efficiency and economic returns. It is therefore suggested that soil characteristics that influence productivity and yield responses should be taken into consideration before embarking on P fertilizer management.

# REFERENCES

1. Barber SA (1995). Soil nutrient bioavailability: a mechanistic approach. 2nd Edition. John Wiley and Sons, New York.
2. Batiano A, Ayuk E, Mokwunye AU (1995). Long-term evaluation of alternative phosphorus fertilizers for pearl millet production on the sandy Sahelian soils of West Africa semi-arid tropics. In H. Gerner and A.U. Mokwunye (eds.) Use of phosphate rock for sustaining agriculture in West Africa. Miscellaneous Fertilizer Studies 11. International Fertilizer Development Centre, Muscle Shoals, AL. pp. 42-53.
3. Beckett PHT, Webster R (1971). Soils and Fertilizers. Soil Variability. A Review, 34(1): 1-15.
4. Carr PM, Carlson GR, Jacobsen JS, Nielsen GA, Skogley EO (1991). Farming soils, not fields: a strategy for increasing fertilizer profitability. J. Prod. Agric. 4(1): 57-61.

5. Holford ICP (1997). Soil phosphorus, its measurement and its uptake by plants. Aust. J. Soil Res. 35: 227-239.
6. International Institute of Tropical Agriculture (1982). Selected methods for soil and plant analysis, IITA, Ibadan, Nigeria. p. 70.
7. Javid S (1999). Residual effect of phosphate fertilizer measured using the Olsen method in Pakistani soils. PhD diss. Univ. Reading, UK.
8. Kolawole GO, Tian G, Singh BB (2000). Differential response of cowpea lines to Aluminium and Phosphorus application. J. Plant Nutr. 23: 731-740.
9. Torres-Dorante LO, Norbert C, Bernd S, Hans-Werner O (2006). Fertilizer-use efficiency of different inorganic polyphosphate sources: effects on soil P availability and plant P acquisition during early growth of corn. J. Plant Nutr. Soil Sci. 169(4): 509-515.
10. McKenzie RH, Bremer E, Kryzanowski L, Middleton AB, Solberg ED, Heaney D, Coy G, Harapiak J (2003). Yield benefit of phosphorus fertilizer for Wheat, Barley, and Canola in Alberta. Better Crops. 87(4): 15-17.
11. Mokwunye AU, Chien SH, Rhodes E (1986). Phosphorus reaction with tropical African soils. In: Management of Nitrogen and Phosphorus Fertilisers in Sub-Saharan Africa. Eds. Mokwunye AU, Vlek PLG, Martinus Nijhoff Publishers, Dordrecht, the Netherlands. pp. 253- 281.
12. Norman M, Rearsonand C, Searle P (1995). The Ecology of Tropical Food Crops. Cambridge University Press, Cambridge.
13. Orkwor GC, Asadu CLA (1998). Agronomy In: Orkwor GC, Asiedu R, Ekanayake J (eds.) Food Yams. Advances in Research, IITA and NRCRI, Nigeria. pp. 105-142.
14. Sanchez PA, Uehara G (1980). Management considerations for acid soils with high phosphorus fixation capacity. In: Khasawneh FE, Sample EC, Kamprath EJ (eds) Proc. on the role of phosphorus in agriculture, June 1-3 1976. ASA CSSA and SSSA Madison, WI, USA. pp. 471-513.
15. Smyth AJ, Montgomery RF (1962). Soils and land use in central western Nigeria. A publication of the Ministry of Agricultural Resources. Government Press, Ibadan. Nigeria. p. 265.

16. Van Ray B, Van Diest A (1979). Utilization of phosphate from different sources by six plant species. Plant Soil, 51(4): 577.589.
17. Zapata F (1986) Agronomic evaluation of rock phosphates for direct application by means of radioisotope techniques In IAEA Experimental Guidelines Vienna, Austria: IAEA.

# Citations

## CHAPTER 1

A. Bah, M. H. A. Husni, C. B. S. Teh, M. Y. Rafii, S. R. Syed Omar, and O. H. Ahmed, "Reducing Runoff Loss of Applied Nutrients in Oil Palm Cultivation Using Controlled-Release Fertilizers," Advances in Agriculture, vol. 2014, Article ID 285387, 9 pages, 2014. doi:10.1155/2014/285387.

## CHAPTER 2

Jacob T. Bushong, D. Brian Arnall, and William R. Raun, "Effect of Pre plant Irrigation, Nitrogen Fertilizer Application Timing, and Phosphorus and Potassium Fertilization on Winter Wheat Grain Yield and

Water Use Efficiency," International Journal of Agronomy, vol. 2014, Article ID 312416, 12 pages, 2014. doi:10.1155/2014/312416.

## CHAPTER 3

Jacob T. Bushong, Eric C. Miller, Jeremiah L. Mullock, D. Brian Arnall, and William R. Raun, "Effect of Irrigation and Preplant Nitrogen Fertilizer Source on Maize in the Southern Great Plains," International Journal of Agronomy, vol. 2014, Article ID 247835, 10 pages, 2014, doi:10.1155/2014/247835.

## CHAPTER 4

Raymond A. Wuana and Felix E. Okieimen, "Heavy Metals in Contaminated Soils: A Review of Sources, Chemistry, Risks and Best Available Strategies for Remediation," ISRN Ecology, vol. 2011, Article ID 402647, 20 pages, 2011. doi:10.5402/2011/402647.

## CHAPTER 5

Richard T. Koenig, Craig G. Cogger, and Andy I. Bary, "Dryland Winter Wheat Yield, Grain Protein, and Soil Nitrogen Responses to Fertilizer and Biosolids Applications,"Applied and Environmental Soil Science, vol. 2011, Article ID 925462, 9 pages, 2011. doi:10.1155/2011/925462.

# CHAPTER 6

Ezekia Svotwa, J. Anxious Masuka, Barbara Maasdorp, and Amon Murwira, "Spectral Indices: In-Season Dry Mass and Yield Relationship of Flue-Cured Tobacco under Different Planting Dates and Fertiliser Levels," ISRN Agronomy, vol. 2013, Article ID 816767, 10 pages, 2013. doi:10.1155/2013/816767.

# CHAPTER 7

G. O. Kolawole, A. O. Olayiwola, O. Ige, G. O. Oyediran, and B. A. Lawal, Evaluation of phosphorus fertilizer rates for maize and sources for cowpea on different soil types in southwestern Nigeria, ISSN 1684–5315.

# Index